H 101

P 105

P 107

Governing Climate Change

Governing Climate Change provides a short and accessible introduction to how climate change is governed by an increasingly diverse range of actors, from civil society and market actors to multilateral development banks, donors and cities. The issue of global climate change has risen to the top of the international political agenda. Despite ongoing contestation about the science informing policy, the economic costs of action, and the allocation of responsibility for addressing the issue within and between nations, it is clear that climate change will continue to be one of the most pressing and challenging issues facing humanity for many years to come.

The book:

- evaluates the role of states and non-state actors in governing climate change at multiple levels of political organization: local, national and global;
- provides a discussion of theoretical debates on climate change governance, moving beyond analytical approaches focused solely on nation-states and international negotiations;
- examines a range of key topical issues in the politics of climate change;
- includes multiple examples from both the North and the Global South.

Providing an interdisciplinary perspective drawing on geography, politics, international relations and development studies, this book is essential reading for all those concerned not only with climate governance but with the future of the environment in general.

Harriet Bulkeley is Reader in the Department of Geography at Durham University, UK.

Peter Newell is Professor of International Development at the University of East Anglia, UK.

Routledge Global Institutions

Edited by Thomas G. Weiss
The CUNY Graduate Center, New York, USA
and Rorden Wilkinson
University of Manchester, UK

About the Series

The "Global Institutions Series" is designed to provide readers with comprehensive, accessible, and informative guides to the history, structure, and activities of key international organizations as well as books that deal with topics of key importance in contemporary global governance. Every volume stands on its own as a thorough and insightful treatment of a particular topic, but the series as a whole contributes to a coherent and complementary portrait of the phenomenon of global institutions at the dawn of the millennium.

Books are written by recognized experts, conform to a similar structure, and cover a range of themes and debates common to the series. These areas of shared concern include the general purpose and rationale for organizations, developments over time, membership, structure, decision-making procedures, and key functions. Moreover, current debates are placed in historical perspective alongside informed analysis and critique. Each book also contains an annotated bibliography and guide to electronic information as well as any annexes appropriate to the subject matter at hand.

The volumes currently published include:

41 Governing Climate Change (2010)
by Harriet Bulkeley (Durham University) and Peter Newell (University of East Anglia)

40 The UN Secretary-General and Secretariat (2010)
Second edition
by Leon Gordenker (Princeton University)

39 Preventive Human Rights Strategies (2010)
by Bertrand G. Ramcharan (Geneva Graduate Institute of International and Development Studies)

38 African Economic Institutions (2010)
by Kwame Akonor (Seton Hall University)

1 The United Nations and Human Rights (2005)
A guide for a new era
by Julie A. Mertus (American University)

Books currently under contract include:

Multilateral Counter-Terrorism
The global politics of cooperation and contestation
by Peter Romaniuk (John Jay College of Criminal Justice, CUNY)

Global Governance, Poverty and Inequality
Edited by Jennifer Clapp (University of Waterloo) and Rorden Wilkinson (University of Manchester)

The International Labour Organization
by Steve Hughes (University of Newcastle) and Nigel Haworth (University of Auckland Business School)

Global Poverty
by David Hulme (University of Manchester)

The Regional Development Banks
Lending with a regional flavor
by Jonathan R. Strand (University of Nevada)

Peacebuilding
From concept to commission
by Robert Jenkins (The CUNY Graduate Center)

Non-Governmental Organizations in Global Politics
by Peter Willetts (City University, London)

Human Security
by Don Hubert (University of Ottawa)

UNESCO
by J. P. Singh (Georgetown University)

Millennium Development Goals (MDGs)
For a people-centered development agenda?
by Sakiko Fukada-Parr (The New School)

UNICEF
by Richard Jolly (University of Sussex)

The Organization of American States (OAS)
by Mônica Herz (Instituto de Relações Internacionais)

FIFA
by Alan Tomlinson (University of Brighton)

International Law, International Relations, and Global Governance
by Charlotte Ku (University of Illinois, College of Law)

Humanitarianism Contested
by Michael Barnett (University of Minnesota) and Thomas G. Weiss
(The CUNY Graduate Center)

Forum on China-Africa Cooperation (FOCAC)
by Ian Taylor (University of St. Andrews)

The Bank for International Settlements
The politics of global financial supervision in the age of high finance
by Kevin Ozgercin (SUNY College at Old Westbury)

International Migration
by Khalid Koser (Geneva Centre for Security Policy)

Global Health Governance
by Sophie Harman (City University, London)

Think Tanks
by James McGann (University of Pennsylvania) and Mary Johnstone Louis
(University of Oxford)

The Council of Europe
by Martyn Bond (University of London)

The United Nations Development Programme (UNDP)
by Stephen Browne (The International Trade Centre, Geneva)

For further information regarding the series, please contact:

Craig Fowlie, Senior Publisher, Politics & International Studies
Taylor & Francis
2 Park Square, Milton Park, Abingdon
Oxon OX14 4RN, UK

+44 (0)207 842 2057 Tel
+44 (0)207 842 2302 Fax

Craig.Fowlie@tandf.co.uk
www.routledge.com

Governing Climate Change

Harriet Bulkeley and Peter Newell

Routledge
Taylor & Francis Group

LONDON AND NEW YORK

First published 2010
by Routledge
2 Park Square, Milton Park, Abingdon, Oxon OX14 4RN

Simultaneously published in the USA and Canada
by Routledge
270 Madison Avenue, New York, NY 10016

Routledge is an imprint of the Taylor & Francis Group, an informa business

© 2010 Harriet Bulkeley and Peter Newell

Typeset in Times New Roman by
Taylor & Francis Books
Printed and bound in Great Britain by
TJ International Ltd, Padstow, Cornwall

British Library Cataloguing in Publication Data
A catalogue record for this book is available from the British Library

Library of Congress Cataloging in Publication Data
A catalog record for this book has been requested

ISBN 978-0-415-46768-1 (hbk)
ISBN 978-0-415-46769-8 (pbk)
ISBN 978-0-203-85829-5 (ebk)

Contents

Illustrations

Figure

Boxes

Foreword

The current volume is the forty-first title—several having already gone into second editions—in a dynamic series on "global institutions." The series strives (and, based on the volumes published to date, succeeds) to provide readers with definitive guides to the most visible aspects of what many of us know as "global governance." Remarkable as it may seem, there exist relatively few books that offer in-depth treatments of prominent global bodies, processes, and associated issues, much less an entire series of concise and complementary volumes. Those that do exist are either out of date, inaccessible to the non-specialist reader, or seek to develop a specialized understanding of particular aspects of an institution or process rather than offer an overall account of its functioning. Similarly, existing books have often been written in highly technical language or have been crafted "in-house" and are notoriously self-serving and narrow.

The advent of electronic media has undoubtedly helped research and teaching by making data and primary documents of international organizations more widely available, but it has also complicated matters. The growing reliance on the Internet and other electronic methods of finding information about key international organizations and processes has served, ironically, to limit the educational and analytical materials to which most readers have ready access—namely, books. Public relations documents, raw data, and loosely refereed websites do not make for intelligent analysis. Official publications compete with a vast amount of electronically available information, much of which is suspect because of its ideological or self-promoting slant. Paradoxically, a growing range of purportedly independent websites offering analyses of the activities of particular organizations has emerged, but one inadvertent consequence has been to frustrate access to basic, authoritative, readable, critical, and well-researched texts. The market for such has actually been reduced by the ready availability of varying-quality electronic materials.

For those of us who teach, research, and practice in the area, such limited access to information has been frustrating. We were delighted when Routledge saw the value of a series that bucks this trend and provides key reference points to the most significant global institutions and issues. They are betting that serious students and professionals will want serious analyses. We have assembled a first-rate line-up of authors to address that market. Our intention, then, is to provide one-stop shopping for all readers—students (both undergraduate and postgraduate), negotiators, diplomats, practitioners from non-governmental and intergovernmental organizations, and interested parties alike—seeking information about the most prominent institutional aspects of global governance.

Governing climate change

In the face of incontrovertible evidence, nonetheless widespread skepticism about the impact of human activity on climate change persists.[1] One consequence prevalent among many conservatives in the industrial North is to attempt to turn the focus of attention away from the greenhouse gas emitting activities of European, North American, Japanese, and Australasian households and the vast multinational corporations such as British Petroleum, Amoco, Shell, Mobil, and Rio Tinto towards worries about the activities of the newly industrializing countries generally, and Chinese and Indian industries and consumers in particular. This political (and quite self-serving, as many of those that propagate this view are connected, either ideationally or more directly, with vested interests) sleight of hand has served to complicate further the problems posed by attempting to put together a coherent set of global machineries charged with the task of mitigating, in some way, the extent to which human activity has had an impact upon the global climate.

For all the bluster of climate change critics, the evidence *is* both incontrovertible and harrowing. The 2007 Nobel Peace Prize was shared by former US Vice-President Al Gore for his work on the environment—specifically his film *An Inconvenient Truth*—and the UN's Intergovernmental Panel on Climate Change (IPCC). The latter, which pulled together the world's foremost scientists, had evolved over 20 years. Its synthetic report found that the "[w]arming of the climate system is unequivocal," and continued:

> Eleven of the last twelve years (1995–2006) [have] rank[ed] among the twelve warmest years in the instrumental record of global surface temperature (since 1850) … From 1900 to 2005, precipitation increased significantly in eastern parts of North and South America,

northern Europe and northern and central Asia but declined in the Sahel, the Mediterranean, southern Africa and parts of southern Asia … cold days, cold nights and frosts have become less frequent over most land areas … [and] heat waves have become more frequent over most land areas … [Moreover] [g]lobal GHG emissions due to human activities have grown since pre-industrial times, with an increase of 70% between 1970 and 2004 … [and] the net effect of human activities since 1750 has been one of warming.[2]

The United Nations system has a long history of involvement in addressing the consequences of human activity on the global environment. Since the first UN Conference on the Human Environment in Stockholm in June 1972,[3] the number of UN and related institutions and mechanisms dealing with the global environment generally, and climate change more specifically, has increased markedly. This proliferation has not, however, been a comfortable one. Differing ideas about how to deal with global environmental degradation have combined with diametrically opposed driving ideologies and different organizational remits to generate a constellation of institutions with little coherence and, as a result, limited capacity.[4] One positive outcome has been that the environment has come to be a major concern for even the most reluctant of global institutions, even if navigating the various roles of myriad bodies in this area has become more problematic. As editors, we wanted to offer readers the most complete and comprehensive guide to global governance in this area to complement other works in the series.[5]

Two better people we could not have hoped to find to write a book for us on the governance of climate change. Both Harriet Bulkeley, a Reader in Geography at Durham University, and Peter Newell, Professor of International Development at the University of East Anglia, are renowned specialists in the area of environmental governance. Together they fuse insights from geography (particularly about space and scale) and international relations (focusing on the political economy of environmental governance) to produce the first major treatment of the various mechanisms involved in the (mis)governance of climate change. The book they have produced is first-rate, and it stands as the most complete treatment on the subject so far. Needless to say, we are very pleased to have the book, and Harriet and Peter, in the series.

As always, we look forward to comments from first-time or veteran readers of the Global Institutions Series.

Thomas G. Weiss, The CUNY Graduate Center, New York, USA
Rorden Wilkinson, University of Manchester, UK
December 2009

Acknowledgements

Both authors would like to thank the series editor Rorden Wilkinson for proposing the idea of a book on climate change, and the publishing team at Routledge, and especially Nicola Parkin, for their assistance with producing the book. They would also like to express their gratitude to Tori Henson for research assistance in the preparation of the final manuscript. They gratefully acknowledge the support of the Economic and Social Science Research Council Climate Change Leadership Fellowships which are supporting their work on climate change governance.

Harriet Bulkeley would like to thank Peter Newell for inviting her to be involved with this project, and for being a committed and thought-provoking co-author. She gratefully acknowledges the support of the Leverhulme Trust 2007 Philip Leverhulme Prize, which has provided her with the time and space to work on this book. She is also thankful for the love and support of Peter Matthews, which continues to sustain her work and life.

Peter Newell would first like to like to express his gratitude to Harriet Bulkeley for being a willing, efficient, and engaging co-author. Second, he would like to thank Diana Liverman for creating the time and space for him to work on the book as a James Martin Fellow at the Oxford University Environmental Change Institute. Third, he would like to show his appreciation for the love and support of Lucila throughout this and many other publication ventures.

The book is dedicated to our respective children Elodie and Ana, for whom the question of the effectiveness of climate governance is of much more than academic interest.

Harriet Bulkeley and Peter Newell
April 2009

Abbreviations

3C	Combat Climate Change
ABARE	Australian Bureau of Agriculture and Resource Economics
AGBM	Ad Hoc Group on the Berlin Mandate
AIJ	Activities Implemented Jointly
AOSIS	Alliance of Small Island States
APP	Asia Pacific Partnership on Clean Development and Climate
BP	British Petroleum
BRICS	Brazil, Russia, India, China and South Africa
C40	C40 Climate Leadership Group
CAN	Climate Action Network
CBDP	community-based disaster preparedness
CCB	CCBA standards
CCBA	Climate Community and Biodiversity Alliance
CCI	Clinton Climate Initiative
CCP	Cities for Climate Protection
CCX	Chicago Climate Exchange
CDCF	Community Development Carbon Fund
CDM	Clean Development Mechanism
CDP	Carbon Disclosure Project
CERES	Coalition for Environmentally Responsible Economies
CERs	Certified Emissions Reductions
CIFOR	Center for International Forestry Research
CO_2	carbon dioxide
CO_2e	Cantor CO_2e
COP	Conference of the Parties
CSR	corporate social responsibility
DEFRA	Department for Environment and Rural Affairs
DNV	Det Norske Veritas

EC	European Commission
ECLAC	Economic Commission for Latin America and the Caribbean
ECO	Environmental Conservation Organization
ERT	European Round Table of Industrialists
ETS	Emissions Trading Scheme
EU	European Union
FIELD	Foundation for International Environmental Law and Development
G77	Group of Developing Countries (originally 77 that signed a declaration in 1964)
G8	Group of Eight
GCC	Global Climate Coalition
GDP	gross domestic product
GEF	Global Environment Facility
GHG	greenhouse gas/es
GRILAC	Group of Latin American Countries
HSBC	Hong Kong and Shanghai Banking Corporation
ICC	Inuit Circumpolar Conference
ICCR	Interfaith Centre for Corporate Responsibility
ICLEI	Local Governments for Sustainability
IETA	International Emissions Trading Association
IFC	International Funding Corporation
IIED	International Institute for Environment and Development
IMF	International Monetary Fund
IPCC	Intergovernmental Panel on Climate Change
ISO	International Organization for Standardization
JI	joint implementation
JUSCANZ	Japan, USA, Canada, Australia and New Zealand
LDCs	least developed countries
LDF	local development framework
LULUCF	Land Use, Land-Use Change and Forestry
NEG-ECP	New England Governors and Eastern Canadian Premiers
NGOs	non-governmental organizations
nrg4SD	Network of Regional Governments for Sustainable Development
ODA	official development assistance
OECD	Organization for Economic Cooperation and Development
OPEC	Organization of Petroleum Exporting Countries

PCF	Prototype Carbon Fund
PPP	public-private partnerships
PRA	participatory rapid appraisal
PSA-CABSA	Payments for Carbon, Biodiversity and Agroforestry Services
QELROs	quantifiable emissions limitations and reduction obligations
REDD	reducing emissions from deforestation and forest degradation
RGGI	Regional Greenhouse Gas Initiative
REEEP	Renewable Energy and Energy Efficiency Partnership
SBI	Subsidiary Body on Implementation
SBSTA	Subsidiary Body on Scientific and Technological Advice
SGS	Société Générale de Surveillance
TCG	The Climate Group
TT	Transition Town
UK	United Kingdom
UN	United Nations
UNCED	United Nations Conference on Environment and Development
UNDP	United Nations Development Programme
UNEP	United Nations Environment Programme
UNFCCC	United Nations Framework Convention on Climate Change
UNICE	Union of Industrial Employers' Confederations in Europe
UNIDO	United Nations Industrial Development Organization
USA	United States of America
VCA	Vulnerability and Capacity Assessment
VCS	Voluntary Carbon Standard
WBCSD	World Business Council on Sustainable Development
WMO	World Meteorological Organization
WRI	World Resources Institute
WRM	World Rainforest Movement
WSSD	World Summit on Sustainable Development
WTO	World Trade Organization
WWF	World Wide Fund for Nature

Introduction

Climate change is one of the most pressing scientific and political challenges of our time. With almost daily news reports of climate-related disasters, international meetings, scientific findings, and various forms of protest, it is unsurprising that the issue is both at the top of the international political agenda and also one of significant public interest. Over the past two decades we have witnessed the perhaps paradoxical twin processes of growing scientific certainty about the causes and consequences of climate change, and rising concern that the issue presents an intractable problem for global governance. While the reports from the Intergovernmental Panel on Climate Change (IPCC) have built an increasingly alarming picture of the changes that increased levels of greenhouse gases (GHG) in the atmosphere may precipitate, the international community has been regarded as slow to take action. In this book, we examine the governance of climate change in order to understand why and how the issue is being addressed and by whom, and to explore the challenges which arise from attempting to confront such a serious threat to our collective well-being.

In this introductory chapter, we consider why conventional accounts of international cooperation and efforts to address the issue may fail to provide a sufficient account of the contemporary landscape of climate governance. We develop an alternative approach, which goes beyond existing thinking about global governance, to help us to understand why, how, and by whom climate change is being governed, and with what consequences.

The governance challenge

Before we consider what different theoretical perspectives might have to say about the nature of climate change governance, it is appropriate to look at the problem itself in more detail. In particular, we suggest

that the complexity of governing climate change stems from three related factors: the multiple scales of political decision-making involved;
the fragmented and blurred roles of state and non-state actors; and the
deeply embedded nature of many of the processes that lead to emissions
of GHG in everyday processes of production and consumption.
Turning first to the issue of the scale of the climate change problem,
it is commonly assumed that climate change is a "global problem." If
we take a closer look at this assumption, however, we can see that the
very nature of how "global" is interpreted can lead to radically different understandings of where, and with whom, the challenge of addressing climate change lies. For some, the global nature of climate change
comes from the physical nature of the problem—because GHG emissions know no boundaries; emissions in one place and time contribute
to increasing atmospheric concentrations which in turn will have
impacts across the globe. Because no one country can tackle climate
change alone, addressing climate change needs a global solution—
usually interpreted as cooperation between nation-states—in order to
reduce emissions worldwide and prevent the problem of free-riding, of
some benefiting from the actions of others. This understanding of the
climate change problem has become orthodox, and the challenge of
governing climate change globally is frequently assumed to be that of
agreeing cooperation, normally in the form of an international treaty,
among the relevant nation-states. In effect, this leads to an understanding of climate change as an international problem since states are
the primary participants in international institutions and have the
authority to sign up to international accords. There are, however,
alternative means through which the global nature of climate change
could be understood. For example, one approach would be to consider
the global processes through which emissions of GHG are generated—
flows of production, trade, and consumption, for example—which
would signal a very different geography of responsibility, suggesting
that multinational corporations and consumers had a more significant
role to play in reducing emissions of GHG than the countries in which
particular goods were produced. For example, Greenpeace International produced a study that compared the carbon dioxide (CO_2)
emissions from the burning of fossil fuels by major oil companies with
those of certain countries and found that Shell emits more than Saudi
Arabia, Amoco more than Canada, Mobil more than Australia, and
British Petroleum (BP), Exxon and Texaco more than France, Spain,
and the Netherlands.[1] Hence, if we think of global as a causal rather
than a spatial category (particularly one bounded by the borders of a
nation-state), we are directed to a very different starting point for

thinking about who governs climate change and where that governing takes place.

At the same time, the very framing of climate change as a global problem tends to neglect the other scales of decision-making which shape the trajectories of GHG emissions and the potential to adapt to climate change. Rather than conceiving of climate change as a global problem, many scholars now suggest it needs to be considered as a multilevel problem, in which different levels of decision-making—local, regional, national, and international—as well as new spheres and arenas of governance that cut across such boundaries—are involved in both creating and addressing climate change.[2] Indeed, we show throughout the book how actors operating at all these levels are involved in the every day governance of climate change.

Opening up the framing of climate change as an international problem also leads to questions about the role of nation-states in its governance. The focus of many analyses of climate change has been on what nation-states, collectively and individually, are or are not doing to combat the issue. This is clearly important—nation-states have significant power and influence over many of the processes which contribute to climate change and reduce vulnerability. However, it is increasingly recognized that nation-states are also limited in the degree to which they can directly effect emissions of GHG and the ability of societies and economies to adapt to climate change. While the language of international agreements and national policy documents often suggests that nation-states can act as containers for emissions of GHG—cutting up the global emissions pie into nation-sized pieces, setting targets and conducting emissions inventories—the GHG emitted within the boundaries of a nation-state are shaped by processes and actors operating across national boundaries, and moreover, only partially within the purview of the state itself. As Geoffrey Heal argues,

> carbon dioxide is produced as a result of billions of de-centralized and independent decisions by private households for heating and transportation and by corporations for these and other needs, all outside the government sphere. The government can influence these decisions, but only indirectly through regulations or incentives.[3]

In part, this reflects the changing nature of the state in the current era of globalization and neo-liberal economic reform. So that whereas nation-states may once have been able to exercise sovereign authority, particularly in those sectors where emissions of GHG are concentrated, such as energy, transport and agriculture, it is now more difficult for at least some of them to do so. For example, many energy markets have been liberalized and privatized, leaving the nation-state with little influence over the generation and supply of energy.[4] At the same time, a globalized economy makes many governments wary of introducing policies such as taxation[5] when businesses threatened with such measures can relocate to areas of the world where there is less or virtually no regulation of their emissions. This means that nation-states have to engage in forms of negotiation and cooperation with non-state actors as they are increasingly dependent on the cooperation of stakeholders and communities—often labeled partnership or enabling approaches—for realizing their objectives. Any analytical framework that seeks to understand how climate change is governed therefore needs to recognize the range of actors that are now involved in the processes of governing climate change.

The critical involvement of non-state actors in climate governance partly reflects a third key feature of the issue—the economically and socially embedded nature of the production of GHG emissions. Unlike other global environmental issues, such as ozone depletion, the causes of which were relatively confined to one industrial sector based largely in Europe and North America, or rainforest destruction, which is confined to specific geographic locations, climate change stems from a complex range of processes, cutting across place and scale. Given that the use of energy is at the heart of modern life and central to achieving economic growth, carving out a specific terrain for climate change governance is problematic both in terms of containing the scope of international and national policy, and also because decisions reached in other domains—concerning trade, energy security, and infrastructure planning to name just a few—will have critical implications for the success or otherwise of efforts for governing climate change.

The multi-scale, multi-actor and embedded nature of the issue poses significant challenges for how we understand and analyze climate change governance. In this book, we develop a new approach to address these challenges. First, we seek to move away from the idea that climate change is exclusively a global issue—an emphasis which implies a narrow focus upon public international policy—and instead to understand the shifting terrain of climate governance across different scales and networks. Second, we shift away from the position that the nation-state is the only or necessarily most important unit of climate politics, to consider the other actors involved, and the ways in which public and private authority operate in climate governance. This involves rethinking where responsibility for addressing climate change might lie, and raises interesting questions about who the relevant actors are in governing climate change, how and why they govern climate change the way they do, and with what implications. Before we set out this approach, let us first look in more detail at existing and conventional approaches to explaining the governance of climate change and the critiques that have emerged.

Climate change: an international problem?

As we outlined above, approaches to understanding climate change as an international problem commonly start with the fact that the atmosphere, in contrast to the world of states, knows no boundaries. Emissions of GHG in one place at one time can have impacts which stretch over time and space in complex ways. Those who have made little or no contribution to the problem will also suffer its consequences, and often it is the most vulnerable who have the least culpability. Managing climate change, from this perspective, entails managing a problem of a resource held in common, over which no one institution or actor has control. In a classic framing of the problem, Andrew Hurrell puts it thus:

> Can a fragmented and often highly conflictual political system made up of over 170 sovereign states and numerous other actors achieve the high (and historically unprecedented) levels of cooperation and policy coordination needed to manage environmental problems on a global scale?[6]

As with other areas in which international cooperation is required, scholars have sought to understand the politics of climate change through the deployment of regime approaches which seek to explain the creation, stability and effectiveness of international institutions and the agreements they oversee (Box I.1).[7] These perspectives proved particularly attractive for understanding the dynamics of international climate change politics, since they respond "to a number of overlapping concerns" that

Box I.1 Defining a regime

International regimes are "social institutions that consist of agreed upon principles, norms, rules, decision-making procedures, and programs that govern the interactions of actors in specific issue areas."

"traditionally characterize the global environmental problematic."[8] These include the desire to regulate states' behavior in order to avoid the so-called tragedy of commons, the need to control tendencies towards free-riding, and the need to respond to the distributive questions arising from the collective response to global environmental challenges. Across a broad body of work concerned with understanding the conditions under which nation-states cooperate to address international problems, there are significant differences in terms of understanding the processes through which regimes are created and operate. Here, we consider three different perspectives and examine the insights that they bring to understanding the process of governing climate change, before considering their shortcomings in the light of the challenges of the climate governance problem raised above.

Understanding regimes

In power-based accounts of regimes, regimes are formed and dominated by a hegemon, the nation-state with the most (economic and military) power, which it can use either to promote cooperation or forestall international agreement.[9] Powerful states can either use their resources and power to create and finance regimes out of an enlightened self-interest in a well-governed international system, such as the United States did with the creation of the World Bank and International Monetary Fund (IMF) at the end of the Second World War, or, by contrast, exercise a veto role by withdrawing their support for a regime or just ignoring it.[10] Here, the design of international institutions is less important than the motivations of the hegemon, and the distribution of power in the international system, in terms of the outcome of the regime. In the case of climate change, as in many other issue areas, scholars have pointed to the United States as a potential hegemon.[11] As we discuss in further detail in Chapter 1, initially the United States took part in the development of the international agreements for addressing climate change— the United Nations Framework Convention on Climate Change (UNFCCC) and the Kyoto Protocol—but withdrew from Kyoto in 2001. On the one hand, the ability of the United States to withdraw from such an agreement on a high-profile issue points to the fragility of regimes and their dependence on the will of the most powerful state.[12] On the other hand, the survival of Kyoto, despite the withdrawal of the United States, and the recent rapprochement between the Obama Administration and the climate change regime, suggests that international institutions are more than the sum of their (powerful) parts. Perhaps more fundamentally, scholars have questioned the basis upon which power is conceived within such approaches. The assumption that interests are predetermined and materially based (on economic wealth or military might) has been challenged in the context of environmental issues that are inherently uncertain and in which interests are difficult to discern at the outset of negotiations.[13] We return to this point below.

A second set of approaches for understanding regimes is also concerned with the role of interests in shaping international cooperation. Unlike power-based approaches, functionalist or interest-based accounts of regimes are concerned with how different institutional designs shape and affect the behavior of nation-states. Interest-based accounts suggest that "regimes are formed when state actors perceive that individual actions with respect to a given issue-area will not promote their interests in the long run. Regimes, here, are seen as the medium used by state actors to reduce vulnerability, opportunism, and uncertainty while stabilizing the expectations needed to promote collective action" over the long term.[14] The focus is on the ability of institutions to reduce transaction costs, share information and enable communication and so build trust and provide a shadow of the future; a sense that short-term engagement will pay long-term dividends. Hence, from this perspective many of the disincentives to cooperate that power-based approaches emphasize are addressed by the ability of institutions to build cooperation through trust, bargaining, and the careful design of institutions which can deter free-riders.

In a third approach to understanding regimes, the emphasis is shifted away from rationalist and interest-based accounts to consider the roles of norms, values and knowledge in shaping the positions adopted by nation-states and the evolution of international institutions. These constructivist accounts focus on "international regimes as a means through which cognitive and normative aspects of the problem in question come to be constructed and learnt, and in turn shape the ways in which states perceive their interests."[15] By opening up the question of how nation-states come to have interests, and how these evolve, constructivist accounts widen the temporal and spatial horizons of regime theory by including pre-negotiation phases of interest development, domestic processes through which interests come to be conceived, and the range of non-state actors involved in developing norms and knowledge about the nature of the climate change problem and how it should be governed.[16] One key aspect of this work has been to point to the role of scientific knowledge in shaping the international politics of climate change through the role of epistemic communities, who "are both politically empowered through their claims to exercise authoritative knowledge and motivated by shared causal and principled beliefs."[17] The IPCC could be regarded as fulfilling the criteria for an epistemic community, with a shared understanding of the causal processes involved in climatic change, normative beliefs (e.g. in a precautionary approach), common tests for the validity of knowledge (e.g. the extensive processes of peer review involved in IPCC reports), and a common policy project based on the need to reduce emissions of GHG.[18] However, whether and how this consensual knowledge has influenced the development of the climate change regime is less clear. While the IPCC was regarded as playing a critical role in the early stages of the development of the regime, once negotiations began their influence can be seen to have diminished. For example, the IPCC has regularly stated that reductions in emissions of GHG in the order of 60 percent of 1990 levels by 2050 are needed in industrialized countries. At Kyoto, a collective target of 5 percent reductions by 2010 was agreed, which is rather far removed from the scientific consensus. As Paterson argues,

> it was clearly not the case that an epistemic consensus neatly produced international co-operation on the climate issue. Instead, it produced resources for policymakers from different countries (or from within different parts of the state within those countries) to advance the positions they preferred—it became another strategic argument at their disposal. Thus, oil producing countries were able to emphasize the uncertainties (even those within the limits of the scientific consensus).[19]

As we discuss in more depth in Chapter 1, rather than a process in which consensual science is unproblematically translated into policy action, the relation between science and climate policy is complex and deeply politicized.[20] This issue points to a wider problem with constructivist approaches. The focus on knowledge and norms can belie the fragile and contested process through which interests are forged, and overlook other important material and political factors that shape the positions taken by different actors, which reduces their explanatory power in relation to the governance of climate change.

Critiques

Regime approaches have provided powerful tools for understanding and analyzing the politics of climate change at the global level. We have shown how different accounts of regimes have demonstrated the roles of powerful states, international institutions, non-state actors, and structural factors in shaping the international climate negotiations and policy outcomes. However, we find that regime approaches are also limited in their scope, particularly with respect to the key challenges of climate change as an environmental problem—the multiple scales and actors involved, and the close integration between the causes and consequences of climate change, economic development, and everyday life. Here, we highlight four critiques of regime approaches which we suggest restrict its analytical power for understanding the governance of climate change.

The first issue concerns where the governance of climate change is seen to take place. In keeping with orthodox accounts of the global nature of climate change, which we discussed at the start of this chapter, regime approaches tend to assume that the location of climate change as a governance problem is in a discrete political domain which we can label "the global." As a consequence, different levels of governance—notably domestic and international arenas—remain treated for the most part as distinct. While some scholars have sought to understand the interplay, or interaction, between different levels of governance—for example, domestic and international politics—this model of hierarchical, discrete scales underplays both the fluidity of interaction at such frontiers, as well as how many of the processes that cause climate change operate across these boundaries.[21]

The second issue relates to who is regarded as conducting the governance of climate change. Regime approaches continue to take "as given the preeminent status of the nation-states as the key point of reference in seeking to account for the ways issues unfold in the global agenda."[22] While some accounts acknowledge the role of non-state actors, their function is primarily determined with respect to how they have sought to influence nation-states and international institutions. This state-centric approach is, we think, problematic, for it serves to limit our analysis of how non-state actors are involved in climate governance. Furthermore, as an entity the state remains essentially black-boxed—the nature of the state and how it operates is taken for granted so that questions concerning how state institutions vary with historical and geographic context, and of how state power is used in practice are rarely raised. One side effect is the extension of a Western conception of the state, its capacity, and role in society, as a model for states everywhere, ignoring important differences in the nature of the state across the globe.[23] Another is that the state is regarded as a neutral actor with regard to the actors and interests it is meant to regulate; an assumption which overlooks the state's dependence on economic actors, particularly in key sectors, such as energy, which are so central to all other industries, and the structural power that this confers on business actors to shape the context in which states make decisions about which courses of action are desirable and possible.[24]

The third issue concerns how climate governance takes place. In so far as it is considered explicitly at all, climate governance is conceived as a top-down process, implemented through international agreements, national policies, and various forms of market instrument. It is assumed that this happens in a cascade fashion in which decisions and authority flow downwards from one level to the next in a linear way.[25] While different traditions of regime theory offer alternative explanations for such processes—whether they be rationally based or driven by new norms or knowledge—little attention is paid to the actual practices through which policies and measures are achieved and which other actors are implicated in this process, and the implications of this for what is and is not included in the remit of climate governance.

Finally, we suggest that regime approaches have to date failed to provide sufficient insights as to why climate governance is taking place, and indeed to the particular form that it takes. While explanations are offered in terms of the power of individual actors, institutions or norms, such approaches work with a particular concept of power which regards it as a force exercised by some actors over others, neglecting both the structural and generative aspects of power. On the one hand, by focusing on the power of actors and of ideas to produce particular outcomes, regime approaches tend to neglect the ways in which power operates to structure the rules of the game within which such contests occur. This in turn means that questions concerning who is served by particular arrangements or forms of governance, of who wins and who loses and why, are often ignored. On the other hand, regarding power as a discrete property of some actors implies the conceptualization of power in zero-sum terms such that, for example, "an increase in the power of non-state actors is ipso facto defined as a simultaneous reduction on state power and authority."[26] Conceiving of power differently, as a facility to "enable things to happen,"[27] allows us to consider the way in which power is generated through the process of governing itself, and move beyond dualistic accounts of the roles of state and non-state actors in addressing climate change.

Global governance perspectives

In part as a response to concerns about the analytical power of the regime approach, scholars within the discipline of international relations have increasingly turned to the concept of governance as a means of understanding the nature of world affairs. Across the social sciences, the term governance has gained in currency as a means of understanding the changing nature of the state and the proliferation of actors and mechanisms involved in the governing of societies. In its broadest sense, governance "relates to any form of creating or maintaining political order and providing common goods for a given political community on whatever level."[28] The specific variant of this literature concerned with global governance owes much to the work of James Rosenau and his distinction between "government," confined to the world of states, and "governance," regarded as a broader phenomenon:

> Governance occurs on a global scale through both the co-ordination of states and the activities of a vast array of rule systems that exercise authority in the pursuit of goals and that function outside normal national jurisdictions. Some of the systems are formalized, many consist of essentially informal structures, and some are still largely inchoate, but taken together they cumulate to governance on a global scale.[29]

Global governance therefore encompasses the numerous activities which are significant both in establishing international rules and in shaping policy through 'on-the-ground' implementation, even when some such activities originate from actors that, technically speaking, "are not endowed with formal authority."[30] As Rosenau suggests, and in common with other definitions of governance, it is the involvement of non-state actors in the governing of collective affairs in particular, that sets global governance aside from other forms of international relations.

Within this broad church, several authors have identified different strands of thinking.[31] Here, we suggest that two main approaches for understanding the nature and dynamics of global governance can be discerned.[32] The first approach "resides within the realms of regime theory" where the concept of governance "has provided a way of rethinking regimes as enmeshed in broader systems of governance instead of issue areas."[33] Closest in approach and analysis to the constructivist version of regime theory, work in this vein moves beyond examining how non-state actors have influenced the positions of nation-states or international negotiations to examine the significant role of non-state actors in governing climate change within and alongside the regime. Much of this work has focused on the role of nongovernmental organizations (NGOs).[34] Such organizations can adopt "insider" strategies, seeking to shape the governing of issues such as climate change through the provision of advice, knowledge, and the development of policy solutions.[35] Within the international arena, NGOs are seen to act as diplomats and "perform many of the same functions as state delegates: they represent the interests of their constituencies, they engage in information exchange, they negotiate, and they provide policy advice."[36] For example, in her analysis of the role of NGOs in the development of the Kyoto Protocol, Michele Betsill argues that it was the specialized knowledge and expertise of NGOs in the Climate Action Network (CAN) that provided them with leverage in the negotiations. However, she suggests that the influence of NGOs was not confined to the technical arena. While there was little evidence of direct influence, in the form of the uptake of particular targets or proposals made by CAN for the Kyoto Protocol, its presence made a significant difference to the overall outcome by holding the European Union (EU) to its negotiating position of a 15 percent cut in emissions of GHG by 2010 and pressuring Al Gore, then US vice president, to take a flexible approach to target-setting. These were two important ingredients in the achievement of the final outcome of Kyoto which may not have been present without the role of the NGO community.[37] In this manner, NGOs played a role independent of individual nation-states, acting with a degree of autonomous agency in shaping the international regime.

The second approach to understanding global governance moves beyond the international arena "by acknowledging the emergence of autonomous spheres of authority beyond the national/international dichotomy" and by focusing "on the complex interlinkages between different societal actors and governmental institutions."[38] Hence rather than focus solely on the role of non-state actors in shaping international climate institutions, such as the UNFCCC and the related Kyoto Protocol, this body of global governance work considers, for example, global climate governance as comprising of "all purposeful mechanisms and measures aimed at steering social systems towards preventing, mitigating, or adapting to the risks posed by climate change."[39] Scholars have examined the various roles of actors from global civil society,[40] municipal and regional networks,[41] public-private partnerships,[42] and the private sector[43] in the governance of global environmental issues. In Chapters 3, 4 and 5 we examine examples of such approaches with respect to climate change in more detail.

Critiques and prospects

If we return to the core critiques raised with respect to regime approaches—concerning, where, by whom, how and why climate governance takes place—we can see that despite claims to offer an alternative, global governance perspectives share many, if not all, of the same shortcomings.

In terms of where climate governance is seen to take place, perspectives from global governance offer a significant departure from regime approaches. Rather than being confined to the international arena, much of the work conducted under the banner of global governance acknowledges both the multi-level and multi-arena nature of climate governance and specifically seeks to examine how the governing of climate change is taking place beyond the international regime. In this book, we follow this line of argument and seek to examine the multiple sites—from the international arena and national governments, through to transnational networks and private sector projects—in which climate governance takes place.

With regard to the second issue, concerning who governs climate change, there is perhaps less distinction from regime approaches. While both perspectives on global governance extend the range of actors involved in governing climate change, for many, in particular those adopting a regime-based account of global governance, the nation-state remains the dominant force. For example, for Paul Wapner "states remain the main actors in world affairs, and their co-operative efforts to establish regimes remain the essential building blocks of global governance in environmental and other issues."[44] Furthermore, the realms of the state and the non-state are regarded as distinct and separate, leading some scholars to conclude that the rise of non-state actors in global governance must be leading to a decline of the state. This black-boxing of the state (and non-state) spheres is unhelpful, for it fails to account for the relations between state and non-state actors, and the relation between public and private authority.

Work in the tradition of international political economy offers a starting point for crossing such divides, bringing into focus the nature of the state, and arguing for its intimate relation with private actors and civil society.[45] Such approaches move away from the notion that the state is either a unitary actor or occupies an autonomous social sphere. Instead, the state is conceived as a "dynamic system of strategic selectivity"[46] through which the hegemony of one social group, or historic bloc, is attained and maintained.[47] From this perspective, sometimes referred as a neo-Gramscian approach, because of insights derived from the writings of Antonio Gramsci about the practice of hegemony, the focus is the coalitions and alliances that different actors construct to secure, challenge or maintain their power.[48] A strong emphasis is placed on the links between production, power, and governance so that processes of governing which ultimately seek to regulate and change forms of production, as climate governance does, have to negotiate with those whose have control over that production and their resulting power and influence.[49] This approach opens up the constitution of state/non-state relations, and requires that we "think through the many ways in which power is consolidated in institutions normally considered outside of the state."[50] Such perspectives allow us to ask important questions concerning who is governing on whose behalf, and who is excluded from such processes. In the remainder of this book, we seek to deploy perspectives from political economy to help us understand who governs climate change, and with what implications.

Like regime approaches, perspectives on global governance have also been criticized for failing to pay due attention to how governing occurs. While studies of global governance are concerned with the processes through which it takes place, Sending and Neumann suggest that

> their ontology and concomitant analytical tools are not equipped to grasp the content of the processes of governance itself. Rather, studies of global governance typically focus on the changing roles and power of state and nonstate actors, and on resulting changes in the institutionalization of political authority.[51]

As a result of this focus on the shifts in power and the institutionalization of authority, questions concerning how governance is achieved in practical terms have been relatively neglected. This is an important omission for, as scholars of governmentality observe, it is through such processes that certain discourses or frames of governing are made material and "certain identities and action-orientations are defined as appropriate and normal."[52] Drawing on these insights, in this book we seek to develop an understanding of how climate governance takes place, and with what implications for how the "problem" of climate change and "its solutions" are normalized within the governance arena.

A combined lack of engagement with how and by whom climate governance is being undertaken contributes to the neglect of questions concerning why it is conducted within the literatures on global governance. In the main, key concepts concerning power and authority are similar to those used within regime approaches, such that both structural and facilitative notions of power have not been used as a means of understanding why, and with what implications, the governing of climate change is taking particular forms. In the remainder of this book, we draw on these alternative notions of power in order to understand the drivers and consequences of climate governance.

Outline

Our brief discussion of regime approaches and perspectives on global governance has demonstrated that both approaches have shortcomings when it comes to explaining where, by whom, how and why climate governance is being conducted. We have also identified prospects for taking forward the considerable insights generated in these bodies of work, acknowledging the multi-scale, multi-actor and embedded nature of the climate governance challenge, by opening up our analysis of the sites of climate governance, considering the relation between state and non-state actors in terms of who is governing climate change, examining the processes through which climate governance is being achieved, and using alternative concepts of power in order to understand why, and with what consequences, climate governance is taking place. We draw on these themes throughout the remainder of the book, and return to consider their implications in our conclusions in Chapter 6.

In Chapter 1 we provide a brief outline of the history of the emergence of climate change as a global governance problem, examining the key roles played by science, North–South politics, and the growing use of market instruments in structuring the international climate regime and the possibilities for addressing the issue. Chapter 3 focuses in more detail on the conflicts that have arisen between countries in the North and South in addressing climate change, and considers in particular what the implications of climate governance have been for issues of equity. Together, these chapters provide a picture of the emergence and development of climate change as an international and national political issue and of the resulting consequences across different policy sectors and different places.

Chapters 3, 4 and 5 examine the emergence of climate governance in a range of political sites beyond the regime and nation-state. Chapter 3 considers the growing phenomenon of transnational networks and partnerships for addressing climate change, examining how such processes of governing take place and assessing their implications in terms of the effectiveness and equity of climate governance. Chapter 4 focuses on the emergence of "community-based" climate governance, examining its connections to international and national policy arenas and the possibilities for self-governing approaches for addressing the issue. Chapter 5 explores the role of private actors in climate governance, considering their significance in both shaping the positions of nation-states and the international regime, and in performing governance functions in their own right. These chapters demonstrate that there is a wealth of climate governance taking place outside of the formal sites of international and national climate politics, which both carry the promise of innovative solutions and also raise considerable challenges in terms of accountability, equity, and effectiveness. In Chapter 6 we conclude by summarizing our main findings and consider the key themes that have emerged through the book and what challenges they raise conceptually and politically.

1 Governing climate change
A brief history

Although we often assume that the high profile that climate change now enjoys means that it is a new political issue, in fact it has a much longer legacy. In this chapter we provide a brief history of the politics of climate change. We analyze the key issues and conflicts in the international negotiations, from the negotiation of the 1992 United Nations Framework Convention on Climate Change (UNFCCC) through to the Kyoto Protocol in 1997 and up to the present state of negotiations towards the Copenhagen summit in 2009. The discussion is organized around three features of climate governance which have characterized this period: the role of science and the scientific community in the governance of climate change, the role of North–South politics, and finally the increasing marketization of climate governance. Before examining these key issues in turn, we provide a brief overview of how climate policy is made at the international level.

Making policy on climate change

Actors and institutions

The international negotiations on climate change are organized around a number of key actors, institutions, and decision-making processes that it is necessary to understand to follow the discussion that follows. In terms of international organizations, three institutions are critical to the process of negotiating climate change policy. First, there is the secretariat of the UNFCCC, based in Bonn since 1996, which organizes and oversees the negotiations, prepares the necessary documentation and is responsible for overseeing reporting of emissions profiles and projects funded through the Kyoto Protocol. Guided by the parties to the Convention, it provides organizational support and technical expertise to the negotiations and institutions, and facilitates the flow of authoritative information on the implementation of the Convention. It has a key and often underestimated role to play in shaping the outcomes of the negotiations.[1] It has an executive secretary who has the responsibility of trying to guide the negotiations towards a successful conclusion. Second, there is the Conference of the Parties (COP) to the UNFCCC and Kyoto Protocol, which meets annually to review progress on commitments contained in those treaties and to update them in the light of the latest scientific advice. This is the ultimate decision-making body in the climate negotiations. Third, there are the Subsidiary Bodies on Implementation (SBI) and Scientific and Technological Advice (SBSTA) and the Ad Hoc Working Groups that take forward negotiations on specific issues which the COP ultimately has to approve. For example, at the moment there is an Ad Hoc Working Group on Further Commitments for Annex 1 parties under the Kyoto Protocol.

In order to shape this process, governments often organize themselves into blocs and negotiating coalitions to enhance their influence and to advance common agendas. These key coalitions and negotiating blocs emerged early on in the negotiations, but have evolved significantly since then as the issues have changed and their levels of economic development have dramatically altered.[2] At one end of the spectrum the Organization of Petroleum Exporting Countries (OPEC) grouping quickly emerged as the coalition of states most hostile to action on climate change. With revenues almost entirely dependent on the export of oil, that opposition was unsurprising. This bloc affected the pace and course of the negotiations, with calls for greater scientific certainty before action could take place, the formation of alliances with businesses opposed to action,[3] and the use of wrecking tactics such as the call for compensation for loss of oil revenues in response to the call from many low-lying developing countries for economic compensation for impacts suffered as a result of climate change.

At the other end of the spectrum the Alliance of Small Island States (AOSIS), a coalition of island states most vulnerable to the effects of sea-level rise, has been the most strident of the negotiating coalitions in its demands for far-reaching and stringent emissions reductions targets. In 1995 it proposed its own protocol to the agreement mandating a 20 percent cut in 1990 emissions by 2005. The AOSIS group works closely with the London-based legal group Foundation for International Environmental Law and Development (FIELD) that provides legal advice on the negotiating text. Indeed, FIELD was attributed a key role in drafting the AOSIS protocol proposal, suggesting the fragility of rigid distinctions about who exercises power and authority in the governance of climate change.[4]

In between these two polarities lay the G77 of less developed countries + China coalition that emphasized the North's primary contribution to the problem of climate change and sought to deflect calls for the South to make commitments and to ensure that funds committed to achieve the Convention's goals were genuinely additional to existing money for aid. The G77, which as we will see below is now less cohesive, continues to provide a platform for shared concerns about climate change policy. The European Union, meanwhile, has been keen to see a stronger agreement, while the United States, particularly during the administrations of presidents George H. W. Bush and George W. Bush, was resolutely opposed to legally binding cuts in GHG. Japan has adopted a position between these two as host to the summit that produced the Kyoto Protocol, but has often been a reluctant leader because of high levels of industry pressure to not over-commit. Both the United States and Japan were part of the JUSCANZ grouping (Japan, United States, Canada, Australia, and New Zealand) which argued for maximum flexibility in how countries are expected to meet their commitments.

Alongside the formal negotiations organized in plenary sessions and working groups that meet in parallel to discuss specific issues, a bewildering array of non-governmental, business and other organizations are registered to participate in the process. Though they do not have formal voting rights, they are allowed to make interventions and are often admitted onto government delegations where they have access to all the meetings taking place. In many ways, these actors are non-governmental "diplomats" that perform many of the same functions as state delegates: representing the interests of their constituencies, engaging in information exchange, negotiating, and providing policy advice.[5]

We can see, therefore, that the process of making climate policy involves international organizations and institutional structures established for this purpose, coalitions and blocs of state actors, and a range of non-state actors who have sought to influence the process of negotiation in a variety of ways. Before turning to discuss particular aspects of this process in detail, we first consider the main policy milestones that have shaped the current state of international climate policy.

Climate change policy milestones

The governance of climate change as a global political issue has progressed from being a cause for concern among a growing number of scientists to gaining recognition as an issue deserving of a collective global effort orchestrated by the United Nations (UN) (Box 1.1). Over time there has been a deepening of cooperation and a firming-up of obligations to act; a process common to many international negotiations on the environment where a general agreement identifies the need for action and a subsequent protocol contains concrete, legally binding emissions reductions commitments. What is also notable, a theme to which we return below, is the increasing use of market or flexible mechanisms to achieve emissions reductions.

Box 1.1 The global governance of climate change: a short chronology

1988   World Conference on the Changing Atmosphere: politicians and scientists conclude that "humanity is conducting an unintended, uncontrolled, globally pervasive experiment whose ultimate consequences could be second only to nuclear war." The conference recommends reducing CO_2 emissions by 20 percent by 2005.

1990   IPCC publishes its *First Assessment Report*.

1991   The Intergovernmental Negotiating Committee is set up to oversee negotiations towards an international agreement.

1992   154 countries sign the UNFCCC at the United Nations Conference on Environment and Development in Rio, which aims to stabilize emissions at 1990 levels by the year 2000 as part of an overall goal to stabilize GHG "concentrations in the atmosphere at a level that would prevent dangerous interference with the climate system."[1]

1994   The UNFCCC enters into force on 21 March.

1995   The first COP agrees in Berlin that binding commitments by industrialized countries are required to reduce emissions.

1995   The IPCC publishes its *Second Assessment Report*, which establishes that "The balance of evidence suggests a discernible human influence on global climate."[2]

1996   The second COP in Geneva sees the United States agree to legally binding targets to reduce emissions as long as emissions trading is included in an agreement.

Box continued on next page.

1997 More than 150 countries sign the Kyoto Protocol which binds 38 industrialized (Annex 1) countries to reduce GHG emissions by an average of 5.2 percent below 1990 levels during the period 2008–12.

2000 The negotiations at the sixth COP in The Hague collapse amid disagreements, principally between the United States and Europe, about the use of the Kyoto Protocol's flexibility mechanisms.

2001 US president George W. Bush announces that his country is to withdraw from the Kyoto Protocol.

2001 In Marrakesh the final elements of the Kyoto Protocol are worked out, particularly the rules and procedures by which the flexible mechanisms will operate.

2005 On 16 February the Kyoto Protocol becomes law after Russian ratification pushes the emissions of ratified Annex 1 countries over the 55 percent mark.

2004 The Buenos Aires Programme of Work on Adaptation and Response Measures is agreed at COP 10.

2005 The first Meeting of the Parties to the Kyoto Protocol takes place in Montreal at COP 11.

2006 At the Second Meeting of the Parties (COP 12), the Nairobi Work Programme on Adaptation and the Nairobi Framework on Capacity-Building for the CDM are agreed.

2007 The IPCC publishes its *Fourth Assessment Report*.

2007 At COP 13 the Bali Action Plan is agreed, which calls for a long-term goal for emissions reductions; measurable, reportable, verifiable mitigation commitments including nationally appropriate mitigation actions by LDCs; enhanced adaptation, action on technology development and transfer, and financial resources and investment to support the above.

2009 The COP 15 takes place in Copenhagen amid concerns that time is running out to create an international agreement that could enter into force by 2012, when the current Kyoto Protocol target period ends.

Notes:
1 Article 2 of the UNFCCC (1992): http://unfccc.int/resource/docs/convkp/conv eng.pdf
2 IPCC, *Second Assessment Report* (Cambridge: Cambridge University Press, 1995).

The UNFCCC was agreed at the United Nations Conference on Environment and Development (UNCED) summit in Rio in 1992. As the first major milestone in the history of climate diplomacy, the UNFCCC provided a framework for global action on the issue. It sought to emulate the apparent success of the ozone regime, which first produced the Vienna Convention establishing the nature of the problem and the basis for action on it, and subsequently the Montreal Protocol which agreed a phase-out of the most damaging ozone-depleting chemicals. Given the sharp differences of opinion described above and the relative lack of momentum behind the issue at the time, the fact the UNFCCC was agreed at all can be considered a considerable achievement. The agreement set the goal of "avoiding dangerous interference in the climate system," defined as aiming to stabilize concentrations of GHG in the atmosphere, and listed some of the policies and measures that countries might adopt to achieve that end. Acknowledging the vast differences in contributions to the problem, the Convention established the principle of "common but differentiated responsibility"[6] and recognized that developing countries were not yet in a position to assume their own obligations. Efforts they could make towards tackling the issue were made dependent on the receipt of aid and technology transfer from Northern countries that were meant to be "additional" to existing aid budgets.

Attention then turned to how to realize the general nature of the commitments contained in the UNFCCC. With scientific assessments of the severity of climate change becoming increasingly common and growing awareness of the inadequacy of existing policy responses, momentum built for a follow-up to the Convention.[7] The 1995 Berlin Mandate at the first COP sought to promote Quantifiable Emissions Limitations and Reduction Obligations (QELROs), and negotiations thus began towards a protocol which would set legally binding targets to reduce GHG emissions. The Kyoto Protocol concluded in 1997 was the outcome of this. Signed by more than 150 countries, it binds 38 industrialized (Annex 1) countries to reduce GHG emissions by an average of 5.2 percent below 1990 levels during the period 2008–12 (see Box 1.2). It fixes differentiated targets for industrialized countries, while setting in train a process to further elaborate joint implementation schemes, set up an emissions trading scheme (ETS) and create a Clean Development Mechanism (CDM). We discuss these further below.

The process for finalizing the rules and operational details of the Protocol was agreed at COP 4 in 1998 as part of the Buenos Aires Plan of Action. In November 2000 parties met in The Hague at COP 6 to try and complete these negotiations, but failed to do so amid a growing rift between the European Union and United States in particular.[8] Having been party to the negotiations and lobbied hard for the inclusion of market-based mechanisms which would allow industrialized countries maximum flexibility, in 2001 the United States then walked away from the Kyoto Protocol. We will see below that part of the United States' refusal to ratify Kyoto was because its economic competitors in the developing world were not required to reduce their emissions. Without the involvement of the United States, many assumed the inevitable demise of the Kyoto Protocol. If the largest contributor to the problem and most powerful economy in the world was not on board, what incentive was there for others to sign up? In fact, the absence of the United States served to galvanize the European Union and G77 + China into further action, and with the Russian ratification of the Kyoto Protocol in 2005 it entered into force.

Subsequent negotiations have focused on detailed issues concerning the implementation and enforcement of Kyoto and, increasingly, what might come in its place as the end of the implementation period (2012) draws ever closer. At COP 7 the Marrakesh Accords were agreed, which established the rules and procedures for the operation of the flexible mechanisms, including the CDM, as well as details on reporting and methodologies. Importantly, they also established three new funds: the Least Developed Countries Fund, the Special Climate Change Fund, and the Adaptation Fund. This work was continued through to the Buenos Aires Programme of Work on Adaptation and Response Measures agreed at COP 10 in 2004. This was followed at COP 11 in Montreal with the creation of the Ad Hoc Working Group on Further Commitments for Annex 1 parties under the Kyoto Protocol. At COP 12 in Nairobi, dubbed the "Africa COP," there was significant discussion about financing issues and how to increase the number of CDM projects being hosted by the poorest regions of the world, most notably sub-Saharan Africa. The meeting produced the Nairobi Work Programme on Adaptation, and the Nairobi Framework on Capacity-Building for the CDM.[9] The Bali Action Plan agreed a year later at COP 13 has set the path for negotiations towards Copenhagen, calling for a long-term goal for emissions reductions; measurable, reportable, verifiable mitigation commitments including nationally appropriate mitigation actions by Least Developed Countries (LDCs); as well as enhanced adaptation, action on technology development and transfer and financial resources, and investment to support the above.[10] We will see in Chapter 2 how these complex questions of responsibility, and who pays for action on climate change, have come to dominate the negotiations.

Box 1.2 The Kyoto Protocol in brief[1]

Commitments

- Industrialized countries are required to reduce their collective emissions of GHG by an average of 5.2 percent below 1990 levels in the commitment period 2008–12.[2]
- The USA must reduce its emissions by an average of 7 percent; Japan by an average of 6 percent and the EU by an average of 8 percent. Other industrialized countries are permitted small increases, while others are obliged only to freeze their emissions.
- Developed countries are obliged to provide:

  - "new and additional financial resources to meet the agreed full costs incurred by developing country parties in advancing the implementation of existing commitments."
  - "such financial resources, including transfer of technology, needed by the developing country parties to meet the agreed and full incremental costs of advancing the implementation of existing commitments," and
  - "financial resources for the implementation of Article 10, through bilateral, regional and other multilateral channels" which developing country parties can avail of.

Instruments

- Clean Development Mechanism—The aim of this body is to assist developing countries in achieving sustainable development and at the same time to help developed countries "in achieving compliance with their quantified emission limitation and reduction commitments." In effect its purpose is to oversee the implementation of projects funded by developed states wanting to accrue credits for emissions achieved overseas. Participation is voluntary and procedures and modalities for auditing and verifying projects were worked out later in the negotiations. Reduction credits are certified by the CDM to ensure that projects add value to savings that would have been made in their absence (Article 12). In addition, a "share of the proceeds from the certified activities

Box continued on next page.

is used to cover administrative expenses as well as to assist the developing country parties that are particularly vulnerable to the effects of climate change to meet the costs of adaptation."

- Joint Implementation/Actions Implemented Jointly—These activities have to be "additional to any that would otherwise occur" and "supplemental to domestic actions." Scope is provided to include "verifiable changes in stocks of sinks" in parties' assessment of their net GHG emissions (Article 6).
- Emissions Trading (Article 17).
- Implementation is via national reports overseen by teams of experts nominated by the parties.

Notes:
1 Peter Newell, "Who CoPed Out at Kyoto? An Assessment of the Third Conference of the Parties to the Framework Convention on Climate Change," *Environmental Politics* 7, no. 2 (2008): 153–159.
2 Parties are expected to have demonstrated progress in reaching this target by the year 2005. Cuts in the three important gases (CO_2, CH_4, and NO_2) will be calculated against a base year of 1990, and cuts in the long-lived industrial gases (hydroflurocarbons, perflurocarbons and sulfur hexafluride) can be measured against a base year of either 1990 or 1995.

Climate becomes political: science and climate governance

As we have seen above, the process of making climate policy has been shaped by a complex mix of institutions and actors. One of the key issues has been the interface between science and policy, which provided the first impetus for international agreement and continues to be central to global climate politics.

The greenhouse effect—in which particular atmospheric gases, so-called greenhouse gases, act to increase the amount of energy retained in the earth's atmosphere—was discovered by Joseph Fourier in 1824. Other scientists, such as the physicist John Tyndall, helped to further identify the relative radiative forcing values of the different greenhouse gases before the greenhouse effect was first investigated quantitatively by Svante Arrhenius in 1896. The basic science behind the greenhouse effect is graphically portrayed in the diagram below (Figure 1.1). This is a naturally occurring phenomenon. However, by increasing the proportion of GHG in the atmosphere, human activities can exacerbate this effect, leading to higher globally average temperatures and changes

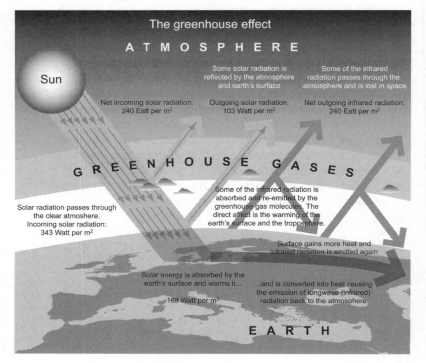

Figure 1.1 The greenhouse effect and climate change.

in the climate. Despite the early discoveries noted above, it was not until the late 1970s, according to climate scientist Bert Bolin, that this "possibility of human-induced change of climate first caught the attention of politicians."[11] During the early 1980s the first international assessment of climate change was convened, supported by the International Council of Scientific Unions and the World Meteorological Organization (WMO), which concluded with the Villach conference in 1985 and an appeal to the international community to address the issue seriously.

The IPCC was then established in 1988 to provide expert input into the climate negotiations, but was to be directed in its mandate by governmental representatives. It was formed by the WMO and the United Nations Environment Programme (UNEP). The IPCC is an expert body made up of the world's leading climate scientists in areas such as oceanography, climatology, meteorology, and economics to produce reports, subject to widespread peer review, on the latest scientific understanding of climate change. Successive reports by the body issued in

1990, 1995, 2001, and 2007 have provided a substantive and consolidated knowledge base about three dimensions of the climate change problem: (i) the science and the basic causal mechanisms, (ii) impacts, and (iii) response strategies: economic, technical and policy implications. Working groups have been created to provide policymakers with the latest research in each of these areas.

Though charged with the responsibility of providing independent and objective advice on the latest scientific understanding of all aspects of the problem of climate change, the IPCC's work has been heavily politicized from the very outset. Science has been a key battleground in the debate about climate change: the severity of the threat, the nature of the causal mechanisms and probable impacts and economic costs associated with taking action. Scientific knowledge is used by all actors in climate governance to advance and defend their position and to confer legitimacy upon it. It is the perception of science as objective and above politics that makes it attractive to actors who believe it provides them with a trump card over the claims of others. However, decisions about whose knowledge counts, who wields authority, and how knowledge is presented and framed are deeply political processes that imply the exercise of power. As Foucault argues, "we should admit ... that power and knowledge directly imply one another; that there is no power relation without the correlative constitution of a field of knowledge, nor any knowledge that does not at the same time presuppose and constitute power relations."[12]

It is unsurprising that the process of accumulating and presenting scientific evidence and advice about the climate change issue has been so political. There is a lot at stake. We see this with the production of policymakers' summaries of the latest IPCC reports, the mostly widely read part of the report. Business, NGOs, and government representatives are allowed to participate in the drafting of these summaries and actively seek to shape their content to reflect their views of the severity of climate change: highlighting uncertainties in the case of those opposed to action or drawing attention to warnings about the potential for dramatic feedback effects by those wanting more stringent near-term action. As former IPCC chairman Bert Bolin reflects: "the Summary for Policy-makers is approved in a plenary session of the respective working groups ... [these] become the occasions when the scientists engaged by the IPCC are confronted with political views and even attempts to clothe special interests in scientific terms."[13]

Recognizing the interwoven nature of power and knowledge challenges the assumption that science interacts with policy in a fairly linear way where governments and international institutions are dependent on experts—sometimes labeled epistemic communities—to advise them on political actions in conditions of uncertainty.[14] In practice, however, the worlds of science and politics overlap. One of the key roles of scientists is to act as "knowledge brokers,"[15] translators of abstract scientific findings into policy messages, intermediaries between lab and law. The fact that scientists are also often dependent on government money for their research has led some to claim that the alleged urgency of the threat posed by climate change has been deliberately exaggerated by a scientific community keen to attract research funding for further work in the area.[16] Once established in a well-resourced and high-profile policy network, so this argument goes, scientists develop their own self-interests in maintaining that position and securing funding for their work. Another source of potent criticism has been that important uncertainties in the science and modeling have been downplayed or ignored altogether. This has been the line of attack of a relatively small but vocal group of so-called climate skeptics who either doubt that global warming results from anthropogenic influences or suggest that the scale of that influence has been overestimated.[17] Given the contentious ground which the IPCC occupies, it is also unsurprising that senior appointments within the IPCC have also been a source of controversy. For example, the US government successfully vetoed the reappointment of Bob Watson as head of the IPCC amid claims he was pursuing a personal agenda which was interpreted at the time as being too outspoken an advocate of climate action.[18]

One of the key issues here is not only that politics get played out through science, but the way science is used politically. While the UNFCCC refers to the prevention of "dangerous anthropogenic interference in the climate system"[19] as the overall goal of the agreement, defining what counts as "dangerous" is a political act, a judgment about risk: how much risk we as a society are willing to take and who bears the consequences. A key battleground in the negotiations towards the UNFCCC was whether sufficient scientific consensus existed to warrant global collective action. Those against more strident forms of action demanded more science, as if that was the key obstacle to action. As the NGO newsletter *ECO* declared at the Geneva climate meeting back in 1990, "all the computer models in the world will not make a Swiss Franc of difference to governments who simply want to sell all the oil the world can be persuaded to buy."[20] Part of the reason the interface between science and policy is so contested is because of the political significance of recommendations that are made for identifying and allocating responsibility. This has been a highly contested issue between developed and developing countries, and it is to this issue that we now turn.

The poverty of climate governance: North–South politics

The second dimension of climate governance which cuts across this history of climate diplomacy is the importance of North–South politics. Climate change in many ways highlights more starkly the North–South divisions which characterize other global environmental threats such as ozone depletion and biodiversity.[21] Early battles about the need to differentiate between the "survival" emissions of the South and the "luxury" emissions of the North[22] have been compounded by claims of carbon colonialism and climate injustice, as we will see in Chapter 2. Underpinning this conflict is the fact that while climate change has been largely caused by the wealthy industrialized parts of the world, it is the least developed areas of the world that will suffer its worst consequences. This makes climate change first and foremost an issue of social justice and equity.

Early discussions of the climate change issue demonstrated a clear polarity of positions between developed countries and developing countries, with the latter arguing that the former, as leading contributors to the problem, were duty-bound to accept responsibility and take action. The language in the UNFCCC embodies this notion of "common but differentiated responsibility,"[23] the idea that everybody has a responsibility to act but some have more responsibility than others. In so far as developing (or non-Annex 1) countries have responsibilities under the Convention, these are conditional on the receipt of aid and technology transfer from more developed countries. Aid and technology transfer provide what economists would refer to as side-payments: inducements to cooperate in tackling a problem to which, historically speaking, they have contributed very little.

Debate has focused on the governance of aid and technology transfer; which institution will oversee these and on whose terms. While developed countries were keen to see the Global Environment Facility (GEF) play a key role, the institution's close association with the World Bank was a source of concern for many developing countries that were distrustful of the organization.[24] Many developing countries also expressed reservations about the levels and types of aid that would be forthcoming. They sought forms of financial assistance that were "additional" to existing aid budgets and not drawn from budget lines targeted for other development priorities.

The presentation of a united front by the G77 on these key issues always disguised a spectrum of diverse national and regional differences. Over time, however, divisions among developing countries have become more apparent and more clearly defined. Increasing emphasis in the global debate on the use of flexible market mechanisms, endorsed by the Kyoto Protocol, forced developing countries to assess how they might benefit from engagement with such mechanisms. Interest in the role of forests as carbon sinks for which credits might be earned and finance provided to developing countries, has drawn biodiverse countries, particularly many Latin American countries, further into the realm of joint actions.[25] At the same time, the newly industrializing powerhouses of China, India, Brazil, Malaysia, Mexico, and South Korea now make their own significant contribution to the problem, raising further questions about how the principle of "common but differentiated responsibilities" should be put into practice.

In this context, there is an ongoing debate about whether, and if so, in what form, developing countries should take on their own emissions reductions commitments. There is a perception among some developed countries that rapidly industrializing competitors will be able to free-ride on the sacrifices made by Annex 1 parties. The related concern is that industries will uproot and relocate to areas of the world not covered by the provisions of the Kyoto Protocol, resulting in "carbon leakage."[26] Increasing recognition of this new geography of responsibility prompted demands from leading polluters—such as the United States and Australia—to draw their competitors into a global regime of regulation and controls on GHG. Most dramatically, at the time of the Kyoto negotiations in 1997, Senator Chuck Hagel drafted the Bryd-Hagel resolution (Senate Resolution 98) which made US acceptance of the terms of the Kyoto Protocol conditional on agreement from leading developing countries to reduce their own emissions. Given the sensitivity about the issue of legally binding emissions reductions obligations for developing countries, this strategy was aimed at causing maximum divisions, stalling progress, and most importantly tying the hands of US negotiators by showing that the Senate would not ratify any deal which did not contain developing country commitments. Indeed, it was this issue in part that led the United States, under George W. Bush, to withdraw from the Kyoto Protocol in 2001.

Besides the issue of responsibility, there is also a clear North–South dimension in terms of vulnerability to the effects of climate change (particularly sea-level rise and changes to agricultural systems). In terms of impacts of climate change, developing countries are in a weaker position to protect themselves from the adverse effects of climate change. Sea defenses, and other means available to wealthier nations to ensure that land is not flooded and that population displacement is not necessary, are not affordable to many developing countries. Their reliance for agricultural production in many low-lying areas that are especially prone to flooding from sea-level rise, makes them particularly vulnerable to the effects of climate change. This is an issue we return to in Chapter 2.

These uneven patterns of responsibility for emissions of GHG and vulnerability to the effects of climate change have profoundly affected efforts to secure global agreements on climate change. The broader historical and contemporary features of the unequal relationship between the developed and developing world run through virtually all aspects of climate governance. Scientists from developing countries are poorly represented in the expert bodies that provide the knowledge base for policy. Poorer groups within these countries are disproportionately vulnerable to the effects of climate change, even if they contribute little to the problem. In many cases, developing country governments lack the capacity to attend, let alone shape and influence, negotiating processes that are heavily dominated by developed countries with the resources to attend all meetings and the legal and scientific capacity to shape developments according to their preferences. The politics of North–South relations have, therefore, both explicitly and implicitly shaped the history of international climate policy, with profound implications for how the issue is addressed. We return to this debate in Chapter 2.

The marketization of climate governance

The third notable feature of climate governance which cuts across this history is the increasing emphasis on market-based solutions. Often referred to as "flexibility mechanisms," the right of countries to meet their obligations through funding emissions reductions elsewhere through the joint implementation of projects or through the buying and selling of permits has been asserted forcefully by the world's most powerful country, the United States. As we can see in Box 1.2, the Kyoto Protocol that was agreed in December 1997 embodied a commitment to constructing such flexible mechanisms.

The support for market-based approaches as the preferred means of climate governance reflects a number of factors. The endorsement of emissions trading in particular built on the apparent effectiveness of other similar schemes, most notably the United States' sulfur dioxide trading scheme. Such approaches reflect the primacy of efficiency as the guiding policy principle of climate governance. In this vein, advocates have argued that it makes no difference where in the world a tonne of carbon is reduced and therefore it makes sense to create mechanisms which allow countries to pay for reductions where it is most cost-effective to do so.

In terms of domestic politics, such mechanisms make emissions reductions easier to sell to electorates wary of burdening domestic industry with additional obligations. Such tools were also seen as a valuable way of bridging some of the North–South conflicts discussed above. They could imply resource transfers to developing countries, potentially on a significant scale, and because emissions are being reduced in these countries they draw them further into the climate regime, pacifying concerns about them free-riding on the sacrifices made by industrialized countries.

That the world of climate governance has become a laboratory for experiments in market-based approaches to regulation is unsurprising, given that climate politics have risen to prominence during an era of heightened neo-liberalism. This broader political context is important in explaining why some solutions are seen as valid, legitimate and plausible, and others are not. It helps us to understand why, for example, command-and-control state-led regulatory approaches are out of favor, or why in a context of globalization and the mobility of capital, it becomes more important than ever to address potential carbon leakage. The emphasis in contemporary neo-liberalism on the creation of (carbon) markets, the allocation of property rights (in this case through allocating permits), the political preference for voluntary partnership-based approaches (such as the Asia Pacific Partnership on Clean Development and Climate (APP)) and voluntary efforts by industry, is clearly apparent in all aspects of contemporary climate governance, as we will see in Chapter 5. The irony, of course, was that having insisted on the inclusion of market-based mechanisms, the United States then refused to ratify the Kyoto Protocol. The European Union, meanwhile, which had initially been hostile to the idea of emissions trading, went on to develop the most advanced emissions trading scheme of its kind, the EU Emissions Trading Scheme.[27]

Despite the prevalence of the idea that markets deliver outcomes more efficiently and effectively than governments, each of the market-based mechanisms mentioned here relies on institutional backing. Governments have to decide who can participate in carbon markets (which sectors and actors) and set limits on the number of permits that are to be traded to create the necessary scarcity to incentivize emissions reductions. Markets, in other words, have to be governed. The 2001 Marrakesh Accords agreed rules, procedures, and modalities for the operation of the CDM, for example (Chapter 2). Furthermore, in order to guarantee the worth of the Certified Emissions Reduction units that are allocated, a range of governance actors have to be enrolled to verify baselines and methodologies, to demonstrate beyond doubt that emission reductions are "additional" (that they would not otherwise have been achieved by other means). Project developers help set up projects and a CDM executive board approves methodologies and projects. Harnessing the market to the goal of climate protection is one thing, but it does not do away with the need for inputs from traditional actors in climate governance, such as states, and international institutions.

Governance issues and challenges

This chapter draws our attention to a number of general features of climate governance. First, though climate change is often talked about as a global governance challenge, those countries whose actions directly determine the shape and effectiveness of the regime are relatively few in number. Deal-brokering unsurprisingly tends to focus on those actors. Traditionally, these have been the United States, the European Union, and Japan. The rising economic power of countries like China, India, Brazil, and South Africa means that they also are increasingly privy to head-to-head closed meetings in which negotiators attempt to establish the basic contours of agreement within which cooperation might be possible. The key deal-brokers in climate governance have changed in a way which reflects shifts in economic power in the global economy. The scope and nature of the negotiations mirrors broader geopolitical and economic shifts in another way too: the support for market mechanisms as the preferred way of governing climate change. Climate governance cannot be understood as separate from the ideology, institutions, and material interests that predominate within the wider global economy in which climate politics exists and with which it seeks to engage.

Second, although traditional approaches to understanding global environmental politics focus on bargaining between nation-states at the international level, they often neglect the importance of domestic politics. Various researchers have revealed the dynamic relationship whereby what happens in the domestic arenas of global powers carries global repercussions, just as what is agreed in global fora reconfigures domestic politics.[28] We have seen in this chapter, particularly in relation to the role of the United States, how global arenas are also a site of domestic politics and domestic politics get played out globally. Analytical distinctions which attempt to neatly separate the two are poorly served by the reality of networks, coalitions, and messy politics that cut across these levels of analysis. Importantly, as we will see in Chapter 2, the effectiveness of traditional climate governance in the form of national regulation and the forms of public international law described here, and the spaces and places where it is said to take place (within states and international institutions), is contingent on the ability and willingness of these governance actors and processes to engage with and seek to transform the everyday practices of climate governance that go by the name of energy policy, development policy, trade, industry, and agricultural policy. These are blind spots or areas of active neglect in the governance of climate change, whose future path will determine whether our collective responses to climate change are up to the challenge.

Third, non-state actors are central to the governance of climate change. Even in the relatively closed world of inter-state diplomacy, non-state actors are on delegations, in the conference rooms, talking to the media and protesting outside the negotiations. The neat distinctions that underpin the official roles and status ascribed to non-state actors describe poorly the reality of actors and individuals that move in and out of these categories, constructing and participating in coalitions and network formations that bypass traditional "levels of analysis," and producing climate governance in their wake. Whether it is the scientific community, business lobbies, or environmental groups, each of these actors plays a role and draws on assets that go beyond what governments and international institutions, the traditional agents in climate politics, can deliver alone.[29] As carbon markets assume increasing importance in the delivery of emissions reductions, business actors in particular play a more important role than ever before. NGOs are increasingly enrolled as project developers, monitors, and watchdogs. Indeed, with so many governance sites beyond the international regime, the UN faces a struggle to retain relevance, and some non-state actors have moved their sights elsewhere. While there are many competing explanations about why the international climate change regime takes the form it does, some of which we discuss in the Introduction (knowledge-based, interest-based and power-based regime theories), the increasing role of non-state actors and the importance of broader (non-regime) economic factors and forces in particular have changed the way we conceive of the whereabouts of the global politics of climate change, how they are conducted and for whom.[30] We return to discuss these issues in more depth in Chapters 3, 4 and 5.

2 Governance for whom?

Equity, justice, and the politics of sustainable development

In this chapter we examine issues of equity and justice in the governance of climate change. As we saw in Chapter 1, from concerns about responsibility for causing the problem and the allocation of the burden of responding to it through to questions of climate adaptation, issues of equity and justice run through every aspect of the governance of climate change. Unsurprisingly, therefore, many activists explicitly invoke the language of "climate justice" when campaigning on these issues.[1]

Picking up from Chapter 1, we place the issue of climate change within the broader context of shifting North–South relations as well as exploring its links to key development concerns around aid, debt, energy and trade, and the ways in which efforts to tackle climate change can be made compatible with the alleviation of poverty. As with other chapters in this book, we suggest that addressing climate change is in reality a sub-set of wider governance processes which shape and at the same time are affected by the nature of climate governance. We look in turn at three key areas where these issues come to the fore: the question of responsibility; the question of who pays for action on mitigation and adaptation; and the question of who bears the costs of actions and inactions. We conclude by summarizing some of the key governance challenges that confront efforts to address issues of justice and equity in relation to climate change.

Whose responsibility? Global warming in an unequal world

Historical and contemporary relations of inequity and injustice form a powerful frame by which poorer countries and peoples make sense of climate change as a political issue.[2] Histories of suspicion, distrust, and inequity are played out in the contemporary politics of climate change, accounting for conflicting notions of whom climate governance should serve and how. Whether it is claims about "carbon colonialism,"[3] ecological debt or climate justice, there is a powerful sense in which climate change has the potential to further aggravate inequalities within and between societies.[4] This sense of injustice derives from the fact that those who have contributed least to the problem of climate change in the past, including most of the world's poor, are those most susceptible to its worst effects now and in the future. Meanwhile, richer countries are better placed to adapt to the climate impacts that they will suffer. Statements by leaders from developing countries illustrate the strength of feeling on the issue. In 2008 President Museveni of Uganda described climate change as "an act of aggression by the rich against the poor."[5] Meanwhile, in March 2003, Ambassador Lionel Hurst, speaking on behalf of the small island states, said:

> the most populous and wealthiest of the world face a moral challenge greater than colonialism or slavery. They are failing in that challenge. Men [sic] have lost reason in the fossil fuel economy … Inhabitants of small islands have not agreed [to be] sacrificial lambs on the altar of the wealth of the rich.[6]

Other critics have focussed on the issue of developing countries being reconstituted as a sink for GHG emissions from developed countries through the "flexibility mechanisms," most critically referred to as "carbon colonialism."[7] Even senior government officials within the climate change negotiations, such as Raúl Estrada-Oyuela, then chairman of the Ad Hoc Group on the Berlin Mandate (AGBM), noted at the time of the Kyoto agreement:

> My reservation was that the CDM is considered a form of joint implementation but I don't understand how a commitment can be jointly implemented if only one of the parties involved is committed to limit emissions and the other party is free from a qualitative point of view. Such disparity has been at the root of every colonization since the time of the Greeks.[8]

The geographies of responsibility have shifted significantly, however, since the early days of the climate regime. While the G77 is able to speak with one voice in demanding further action from the developed countries, it contains among its members the world's largest current contributor to climate change (China) and many other countries whose emissions trajectories will soon match those of the United States, EU, and Japan (for example India, Brazil, and Mexico). At issue here is the extent to which "differentiated responsibilities" should reflect past, current, or future trends of GHG emissions. Any international agreement for the period after 2012 that involves these industrializing countries will have to acknowledge their historically low contributions to the problem even if current levels now match those of developed countries. This is because much of today's climate change has been caused by emissions generated principally by the development of the North. In this sense the North continues to owe a huge "ecological debt" to the South, having consumed a disproportionate share of its global and historical entitlement. Furthermore, developing countries continue to be home to a significant percentage of the world's poor, and their per capita contributions to the problem are dwarfed by those of the United States for example, which is home to just 4 percent of the world's population but is responsible for 20 percent of global emissions, while 136 developing countries are together responsible for 24 percent of those emissions. It becomes clear that the issue of inequity in per capita emissions goes beyond national borders when we consider that the richest 20 percent of the world's population is responsible for over 60 percent of current emissions of GHG.[9] Taking into account the inequity between and within countries concerning per capita contributions to climate change will be a significant challenge for future international agreements.

Beyond the generic framing of common but differentiated responsibility in the UNFCCC, two proposals have been advanced that seek to tackle the thorny issue of per capita emissions and to recognize and reconcile the right to development with the need to dramatically reduce emissions. "Contraction and Convergence" is one such proposal developed by a small London-based NGO called the Global Commons Institute and its charismatic head, the musician Aubrey Meyer. The basic idea which underpins the proposal is that developed countries have to contract their emissions down towards an agreed level which would address the UNFCCC's aim of avoiding dangerous interference in the climate system, while poorer developing countries would be allowed to increase their emissions up to the agreed level. Per capita entitlements would then be bought and sold to harness the market towards this climate stabilization goal. The idea has many opponents, not least the United States and other larger polluters, who insist that permits would have to be allocated based on existing need (a grandfathering system) rather than a per capita basis which would favor developing countries. It is politically and ethically very difficult to argue against the idea, however, that everyone has equal access to the global commons.

A more recent proposal is the Greenhouse Development Rights framework developed by the Stockholm Environment Institute, Christian Aid, and the Heinrich Böll Foundation. It aims to promote the right to development in a carbon-constrained world while holding warming below 2°C, a task which may already seem impossible. Countries' abatement targets would be calculated by the number of people above a development threshold ($20 a day): the "global consuming class" as they refer to it. Responsibility and capacity indicators (gross domestic product (GDP), population, cumulative emissions) including income available after meeting basic needs, are used in the calculations. Such an approach helps to go beyond aggregated national figures of wealth and development, challenging divisions made between countries such as that of "Annex 1" and "non-Annex 1," or "developed" and "developing" countries, while also providing a quantitative handle on how to assess obligations to tackle climate change. Implementation of such an approach could be through taxation paid into a global fund for redistribution, for mitigation and adaptation, or via national mitigation targets as in Kyoto. Questions remain about whether such a proposal is politically realistic, and as such it does not provide a way of bypassing conflicts over how money would be redistributed or by whom or how capacity and responsibility indicators would be agreed. Nonetheless, such proposals show that it is possible to conceive of responsibility for climate change in different ways, with significant implications for the development of climate policy and its ability to address issues of equity.

Such debates are slowly taking shape within the international arena. Although the 2007 Bali Action Plan places emphasis on "nationally appropriate mitigation actions" in less developed (non-Annex 1) countries, the question of how to attribute responsibility for emissions remains hotly contested. In addition to the challenges of historical and future emissions, nation-based or per capita emissions allocations, a further concern is where along global supply networks responsibility for emissions should be allocated—whether the point of production or consumption is the appropriate focus. Many rapidly industrializing countries such as China claim their emissions profile is a by-product of demand for low-cost manufactured goods in developed countries. In other words, they should not be penalized for the fact that many such countries are able to out-source the most carbon and energy-intensive stages of the production process to parts of the world where it is cheapest to do so.

If climate politics could ever have been characterized as at core a conflict between North and South over responsibility, the rise of these issues on the agenda means that the picture is certainly more complex now. The issues around which developing countries once mobilized have changed, as has the political and economic might of those who once relied upon the solidarity of other states as part of the G77 coalition. It is increasingly difficult for the BRICS countries (Brazil, Russia,

India, China, and South Africa) to argue that they should not have obligations to control their emissions of GHG as their profiles match those of countries which are obliged to do so. Likewise, growing reliance upon flexible mechanisms and market-based instruments has served to fragment the opposition of developing countries to involving themselves in emissions reductions projects. The GRILAC (Group of Latin American Countries) has seen opportunities to access funds for forestry projects that might not otherwise be supported. Given this diversity, it is unsurprising that speaking as a united bloc makes increasingly less sense for developing countries in the climate negotiations. Some of the more powerful governments have been drawn into alternative governance fora for reaching agreements on climate change. The preference among India, China, and Brazil is to conclude bilateral deals and engage in partnerships that offer more immediate returns alongside, and some suggest in competition with, Kyoto. These include the Asia-Pacific Partnership on Clean Development and Climate; the Renewable Energy and Energy Efficiency Partnership (REEEP); the Australia-China Partnership; the EU-China Partnership on Climate Change; the US-India bilateral agreement on nuclear energy, and agreements on energy cooperation between the EU and India. As India, China, and Brazil find alternative paths for addressing climate change, we are perhaps left in the international negotiating arena with an "insecure alliance between the unwilling (OPEC) and the powerless (least developed countries) providing few impulses towards increasing G77 flexibility in the post-Kyoto negotiations."[10] The new geographies of responsibility for climate change are, therefore, not only shaping the trajectory of current and future international climate agreements, but can be seen as one of the factors shaping the emergence of a whole range of new governance arenas for addressing climate change, a topic to which we return in Chapter 3.

Who pays?

A second key issue that has shaped the North–South politics of climate change is the concern about who should meet the costs of responding to the issue. Agreeing a fair and equitable allocation of responsibility for acting on climate change requires subsequent agreement about who pays for such actions and how. In governance terms, some of the key conflicts are about which institutions should oversee such flows of funds from North to South and on what terms. As we saw in Chapter 1, the battle over whether the GEF should be the body to oversee aid and technology transfer reflects concerns about the role of the World

Bank, with whom many developing countries have had, and continue to have, difficult relations. A further concern, illustrating how broader North–South politics have shaped climate governance, is where the finance for climate change will come from. Reflecting a deep-seated fear that existing aid budgets might be diverted to fund climate change measures, the obligations for Annex 1 parties explicitly require that they provide "new and additional" financial resources (Box 2.1), though the extent to which this is in fact the case is hotly contested.

Box 2.1 Financing obligations under the UNFCCC[1]

Developed country parties are required to:

- "Provide new and additional financial resources to meet agreed full costs incurred by developing countries parties in communicating information on greenhouse gas inventories and mitigation measures."
- "Provide such financial resources, including the transfer of technology, needed by the developing country parties to meet the agree full incremental costs of implementing GHG mitigation measures."
- "Take into account the need for adequacy and predictability in the flow of funds and the importance of appropriate sharing of burden among the developed country parties."
- "Assist developing country parties that are particularly vulnerable to the adverse effects of climate change in meeting the costs of adaptation to those adverse effects."
- "Take all practicable steps to promote, facilitate and finance, as appropriate, the transfer of, or access to, environmentally sound technologies and know-how" to developing countries to allow them to implement the provisions of the Convention; and
- "Support the development and enhancement of endogenous capacities and technologies of developing country parties."

Developing countries on the other hand:

- "may on a voluntary basis propose projects for financing along with, if possible, an estimate of all incremental costs of the

Box continued on next page.

reduction of emissions and increments of removal of greenhouse gases, as well as an estimate of consequent benefits."
• may obtain financing for Convention-related activities from bilateral, regional or other multilateral sources.

This finance is necessary to help developing countries submit national communications, meet the incremental costs of voluntary projects, combat the adverse impacts of climate change, and undertake adaptive measures deemed necessary.

Note:
1 United Nations, *United Nations Framework Convention on Climate Change* (Bonn, Germany: UNFCCC Secretariat, 1992).

Turning first to questions of institutional oversight, it is clear that multilateral and regional development banks have sought to define for themselves a key role in financing mitigation and adaptation, based on a recognition that climate change can reverse development gains in so many other areas where they work.[11] A key actor here is the World Bank. In addition to being an implementing agency of the GEF, the financing agency for the climate agreements, the World Bank has been involved in a range of climate change initiatives (Box 2.2). The Community Development Carbon Fund (CDCF) launched by the World Bank at the World Summit on Sustainable Development (WSSD) in 2002, with initial capital of $128.6 million, is intended to encourage CDM-like, development-focused project funding. The CDCF provides financial support to small-scale emissions reductions projects through the CDM in the least developed countries and poorest communities within the developing world. A key role for the World Bank and other donors is to create incentives for the private sector to carry some of the burden of funding climate protection. Piloting and demonstrating sustainable energy projects is a key function that banks and development agencies can perform to minimize some of the risks that deter private actors from investing in the poorest areas. The World Bank, the International Funding Corporation (IFC) (the World Bank's private lending arm), and US-based private foundations, for example, have set up a Solar Development Corporation towards this end to provide business development services to local solar entrepreneurs and to provide credit for both solar businesses and purchasers of solar home systems.[12] The World Bank also played a part in advancing the goals of the global climate regime by launching the Prototype Carbon Fund (PCF), with investments from private sector firms, intended to facilitate acquisitions and transfers of credits from the flexible mechanisms created by the Kyoto Protocol. The PCF has been operational since 2000, with a range of governments contributing over $180 million to its financing. The PCF has shareholders from the public and private sectors, and was set up to create carbon as an asset for trading in the marketplace.

Box 2.2 Examples of World Bank carbon finance[1]

BioCarbon Fund

The World Bank has mobilized a fund to demonstrate projects that sequester or conserve carbon in forest and agro-ecosystems. The Fund, a public/private initiative administered by the World Bank, aims to deliver cost-effective emission reductions, while promoting biodiversity conservation and poverty alleviation. The Fund is composed of two tranches: Tranche One started operations in May 2004 and has a total capital of $53.8 million; Tranche Two was operationalized in March 2007 and has a total capital of $38.1 million. Both tranches are closed to new fund participation.

The BioCarbon Fund considers purchasing carbon from a variety of land use and forestry projects; the portfolio includes Afforestation and Reforestation, Reducing Emissions from Deforestation and Degradation, and is exploring innovative approaches to agricultural carbon.

Carbon Partnership Facility

The new proposed Carbon Partnership Facility is designed to develop emission reductions and support their purchase over long periods after 2012. Its objective and business model are based on the need to prepare large-scale, potentially risky investments with long lead times, which require durable partnerships between buyers and sellers. It is also based on the need to support long-term investments in an uncertain market environment, possibly spanning several market cycles. "Learning by doing" approaches will

Box continued on next page.

be an essential aspect of the Carbon Partnership Facility as it moves from individual projects to programmatic approaches, including methodologies needed for such approaches.

Umbrella Carbon Facility

The Umbrella Carbon Facility is an aggregating facility to pool funds for the purchase of emissions reductions from large projects. The facility would purchase GHG emission reductions from CDM and Joint Implementation projects. The purchases would be made on behalf of interested public and private entities that will have contributed to the facility as they strive to meet their commitments under the Kyoto Protocol or other international regulatory systems (such as the EU's ETS), including contributions from the World Bank's other carbon funds. The proposed facility would have multiple tranches, each tranche buying emission reductions from one or more individual projects (such as one HFC23 destruction facility) or a program (such as a program for coal thermal power sector repowering by a public sector financial institution in India).

Note:
1 World Bank, *Carbon Funds and Facilities*, http://go.worldbank.org/51X7CH8VN0

However, while some parts of the World Bank are promoting action on climate change, many other policies pursued by the same institution—notably energy market deregulation—do not take their impact on climate change into account. For example, the World Bank concedes that "unregulated electricity markets are likely to put renewable energy technologies at a disadvantage in the short-run because they favor the cheapest energy as determined purely by price, but do not capture environmental and social externalities."[13] Moreover, one report found that during the past three years, less than 30 percent of the World Bank's lending to the energy sector has integrated climate considerations into project decision-making. As late as 2007, more than 50 percent of the World Bank's $1.8 billion energy-sector portfolio did not include climate change considerations at all.[14] In 2006 the World Bank raised its energy sector commitments from $2.8 to $4.4 billion. But while oil, gas, and power sector commitments account for 77 percent of the total energy sector program, "new renewables"[15] account for only 5 percent.[16] On this basis, it is clear that climate change has yet to be mainstreamed in the activities of the World Bank.

Whatever their limitations may be, however, development banks do have a key role to play in supporting clean energy transitions in the developing world. Energy is clearly pivotal to development, and meeting the development needs of the majority of the world's people in a carbon-constrained world presents a global challenge of staggering proportions. Today, 1.6 billion people are without electricity. Electricity demand in developing countries is projected to increase three to five times over the next 30 years,[17] and 57 percent of future power sector investment will occur in developing countries.[18] Without a significant change of course, most of this will be fossil-fuel based electricity production that will exacerbate climate change. A recent UN report notes that of the "substantial shifts in investment patterns" required to mitigate climate change, "half of these should occur in developing countries which will require incentives and support for policy formulation and implementation."[19] Effective governance of these broader processes will clearly be key to the success of any efforts at climate change governance.

In this context, the second issue concerning who pays is how to generate new sources of funding. One approach is to expand the amount of private investment in addressing climate change. A UNFCCC report on investment and financial flows to address climate change argues that the "carbon market ... would have to be significantly expanded to address needs for additional investment and financial flows."[20] Generating more money from the existing system is not as unproblematic as it may appear. If the wealth to be redistributed by banks and donors derives from unsustainable climate change accelerating activities (either their own or those of others), it is detrimental to the purpose of tackling climate change. In short, demands for further money for action on climate change, whether mitigation or adaptation, imply enhanced production and consumption of fossil fuels, given their centrality to the way in which wealth is currently generated. A good example of this contradiction is the current proposal to increase funds for adaptation by imposing a tax on aviation, effectively tying the ability of the poorest to survive the effects of climate change to increased flying by the world's wealthiest citizens, an act which itself exacerbates the vulnerability of the poor to climate change.

Another approach has been to channel government funds for mitigating climate change to climate and development benefits in developing countries. In line with this argument, the CDM of the Kyoto Protocol is often highlighted as a great development opportunity for developing countries to use climate-related funds to invest in projects and technologies that benefit the poor. The projects it approves are meant to capture social and environmental benefits and intended to be additional to projects that would have been funded anyway. However, in reality, investment has tended to flow to larger developing countries such as China and India—that between them are currently home to over half of all registered CDM projects—and not, for example, to Africa, where less than 2 percent of projects are currently registered. There has also been a neglect of broader sustainable development benefits flowing from the projects, even if they are successful at reducing the growth of GHG emissions, partly because those benefits are not monetarized in the way emissions reductions are through the allocation of Certified Emissions Reductions (CERs).[21] In governance terms, there are also issues of limited capacity to manage project demand in some countries, as well as to oversee and implement the projects in a way which protects not just the interests of the investor but also the communities hosting projects.[22] We discuss these issues in further depth in Chapters 5 and 6. For the moment, it is sufficient to note that the debate concerning where additional funds for addressing climate change should come from is far from being resolved in terms of either policy or practice, with significant concern expressed that addressing climate change may serve to divert much needed resources away from other environment and development agendas, will be tokenistic at best, and may serve to promote particular solutions to the climate change problem favored by the developed world rather than those developing countries in receipt of climate finance.

There is also the danger of trying to create islands of formal climate governance in a sea of unregulated, ungoverned flows of trade and finance, unguided by the imperative of addressing climate change. The CDM oversees only a fraction of the global flows of public and private investment that need to be steered towards the construction of a low carbon economy. In overall terms only a small percentage of trade, aid, production and finance is governed by public bodies charged with tackling climate change. Official development assistance (ODA) funds are currently less than 1 percent of investment globally.[23] What this means in practice is identifying a series of policies, strategies and interventions which are able to steer financial flows, public and private, to where they are most needed but in ways that are consistent with the goal of reducing GHG emissions. For those countries most integrated within the global trading system (Organization for Economic Cooperation and Development (OECD) plus BRICS), an agreement on trade in energy services and the use of energy-intensity indicators may make sense. For example, proposals for an energy round in the World Trade Organization (WTO) aimed at meeting the energy needs of the poor through access to services and technologies might help to address poverty and climate change simultaneously. As the WTO puts it:

> Facilitating access to products and services in this area can help improve energy efficiency, reduce greenhouse gas emissions and have a positive impact on air quality, water, soil and natural resources conservation. A successful outcome of the negotiations on environmental goods and services could deliver a triple-win for WTO members: a win for the environment, a win for trade and a win for development.[24]

According to the World Bank, the removal of tariffs for four basic clean energy technologies (solar, wind, clean coal, and efficient lighting) in 18 developing countries with high levels of greenhouse gas emissions would result in trade gains of up to 7 percent. The removal of both tariffs and non-tariff barriers could boost trade by as much as 13 percent.[25] For others, such as countries in sub-Saharan Africa, less well integrated into the global economy and more aid-dependent, important support can be provided by donors to enable clean energy transitions. The World Bank and regional development banks, meanwhile, can play a key role in screening public and private flows going into countries that are already attractive investment locations, as well as provide inducements that reduce the risk of investors entering new markets in parts of Asia, Africa, and Latin America that have not received such flows to date.

Who bears the costs?

In justice and equity terms it is necessary to relate the question of who pays for action in financial terms to the question of who bears the costs of climate change, not just in terms of biophysical impacts, but also the human and social costs of how the issue is currently governed or "ungoverned."

The findings contained in the 2007 IPCC report on the impacts of climate change demonstrate quite clearly that poorer and marginalized communities in drought-prone areas, those experiencing water scarcity, or those whose livelihoods depend on agriculture, will be the worst affected and have the least capacity to adapt to climate change. As IPCC lead author Neil Adger notes, "the impacts of climate change are likely to be greater on those countries more dependent on primary sector economic activities [mostly farming], primarily because of the increase in uncertainty on productivity in the primary sectors."[26] As shown in Box 2.3, changes in the supply of water, natural resources, and food bring in their wake enormous social changes and disruptions. Across the developing countries, it is areas of sub-Saharan Africa where poverty is most acute that will be worst affected. Indeed, a multi-donor report on *Poverty and Climate Change* rightly acknowledges that "Climate change is a serious risk to poverty reduction and threatens to undo decades of development efforts."[27]

Given the slow pace of progress of international cooperation on climate change, noted in Chapter 1, and the difficulties of advancing some of the mitigation options discussed above amid conflicts about how and by whom they should be governed, attention has increasingly turned to the issue of adaptation to climate change. Vulnerability to climate change cuts across geographical boundaries, and it is often the poor in

---

Box 2.3 Climate change and poverty[1]

- Food production needs to double to meet the needs of an additional 3 billion people over the next 30 years. Climate change is projected to decrease agricultural productivity in the tropics and sub-tropics.
- One-third of the world's population is susceptible to water scarcity. Populations facing water scarcity will more than double over the next 30 years. Climate change is projected to decrease water availability in many (semi-) arid regions.
- Wood fuel is the main source of fuel for one-third of the world's population. Wood demand is expected to double in the next 50 years. Climate change will make forest management more difficult due to increases in pests and fires.
- Today 1.6 billion people are without electricity. Electricity demand in developing countries will increase by three to five times over the next 30 years. Fossil fuel-based electricity production will exacerbate climate change.

Note:
1 O. Davidson, K. Halsnaes, S. Huq, M. Kok, B. Metz, Y. Sokona and J. Verhagen "The Development and Climate Nexus: The Case of Sub-Saharan Africa," *Climate Policy* 3, no. 1 (2003): 97–113.

all parts of the world that are affected most, raising issues of distributive justice. We saw in the case of Hurricane Katrina how poorer communities of color bear some of the worst devastation inflicted by "natural" disasters. Poorer people often have the least means of adapting their livelihoods and means of survival to sudden change of the sort associated with extreme weather events. Bradley Parks and Timmons Roberts argue, for example, that Hurricane Mitch in Honduras "serves as a parable about uneven vulnerability to global climate change."[28] The Economic Commission for Latin America and the Caribbean (ECLAC) estimated that $5 billion would be necessary to fund reconstruction efforts in Honduras, to say nothing of damage to the banana plantations that were so central to generating export revenues—revenues that fell to one-fifth of their pre-hurricane levels, forcing Dole and Chiquita to lay off 25,000 workers and setting in train a series of other devastating social consequences. Likewise, the torrential rains and tropical cyclones which struck Mozambique and surrounding countries in 1999 left 700 people dead, 1 million people displaced, and extensive losses of land, animals and other means of survival, leaving a reconstruction bill of $700 million.[29] Constructing effective mechanisms of adaptation raises issues not just of distributive, but also of procedural justice.[30]

The concern with issues of procedural justice reflects the fact that the impacts are felt and costs borne by poorer groups in particular, not just as a result of climate change and differential ability to adapt to it, but because of the ways in which mitigation and adaptation policies are designed and implemented. Though the UN climate change negotiations may seem far removed from local land conflicts, the governance of climate change intersects with the governance of other key resources such as water and land. For example, there have been cases of negative impacts on the poor when land is given up for carbon sink purposes, giving rise to accusations of carbon colonialism. For example, a Norwegian company operating in Uganda that leased its lands for a sequestration project is said to have resulted in 8,000 people in 13 villages being evicted.[31] The project in Bukaleba Forestry Reserve was meant to offset GHG emissions of a coal-fired power plant to be built in Norway. International criticism at the time prevented the project from claiming carbon credits to "offset" the power plant emissions, but the project continued and the trees were planted.[32]

It is unlikely and improbable, however, that people will meekly accept their fate as victims of forms of climate governance they consider unjust. As watchdogs, NGOs have played an important role in monitoring public and private activity in the new carbon economy and drawing attention to the inequities it produces or fails to address. SinksWatch, for example, is an initiative of the World Rainforest Movement (WRM) set up in 2001, that monitors the impact of the financing and creation of sinks projects to highlight the threat they pose to forests and other ecosystems, to forest peoples, and to the climate. A particular concern is the exclusion of marginalized groups from their own forest resources once they become the property of a distant carbon trader for whom they represent a valuable investment opportunity. SinksWatch has focused in particular on projects funded by the World Bank PCF. One case in point is the Plantar project, which involves planting 23,100 hectares of eucalyptus tree plantations to produce wood for charcoal, which will then be used in pig iron production instead of coal. The project also claims carbon credits for sequestration of carbon through the trees planted. Because of the social and environmental impacts of the project, over 50 Brazilian NGOs, movements, churches, and trade unions have been urging PCF investors since March 2003 to refrain from buying carbon credits originating from the project.[33]

In the search for means of protecting poorer groups from the effects of climate change or the effects of poorly conceived forms of climate governance, many affected communities and activities have increasingly turned to the language and tools of human rights. As Kevin Watkins of the United Nations Development Programme (UNDP) said of climate change: "it is about social justice and the human rights of the world's poor and marginalized. Failure to act on climate change would be tantamount to a systematic violation of the human rights of the poor."[34] As a mechanism for seeking redress for rights violations, access to the law is limited for many poorer groups, but there has been a series of legal cases which seek to explore the boundaries of culpability and establish new and challenging legal precedents.[35] One such case involves the Inuit community in North America. On 7 December 2005 the Inuit Circumpolar Conference (ICC) submitted a "Petition to the Inter-American Commission on Human Rights Seeking Relief from Violations Resulting from Global Warming Caused by Acts and Omissions of the United States." The United States was targeted as the world's largest contributor to GHG emissions. Sheila Watt-Cloutier, Inuk woman and chair of the ICC, submitted the petition on behalf of herself, 62 other named individuals "and all Inuit of the Arctic regions of the United States and Canada who have been affected by the impacts of climate change."[36] The petition calls on the Inter-American Commission on Human Rights to investigate the harm caused to the Inuit by global warming, and to declare the United States in violation of rights affirmed in the 1948 American Declaration of the Rights and Duties of Man and other instruments of international law such as the International

Convention on Civil and Political Rights, and the International Convention on Economic, Social and Cultural Rights. Specifically, the petition alleges:

> The impacts of climate change, caused by acts and omissions by the United States, violate the Inuit's fundamental human rights protected by the American Declaration of the Rights and Duties of Man and other international instruments. These include their rights to the benefits of culture, to property, to the preservation of health, life, physical integrity, security and a means of subsistence and to residence, movement, and the inviolability of the home.[37]

The Commission rejected the petition as inadmissible, though reasons for the refusal were not given. Importantly, however, the human rights issues raised by the case were not disputed by the Commission. One positive outcome of the case has been that the Commission invited petitioners to request a public hearing on the matter, which took place on 1 March 2007.

Civil society actors North and South also have a key role to play in debates about equity and justice in climate governance through enhancing the voice and reach of excluded groups. This can take the form of legal groups, such as the UK-based FIELD, helping to redress some of the procedural inequities which mean levels of participation and effective engagement from smaller developing countries in the climate negotiations are low. Outside the formal negotiating arenas of climate governance, it would also include the Climate Justice movement, which draws attention to the disproportionate impact of climate change on the poor. This movement focuses on the way in which climate change has the potential to exacerbate existing social inequalities and draws much of its critique from broader contestations of neo-liberal globalization. For the activists spearheading the movement, Climate Justice means "holding fossil fuel corporations accountable for the central role they play in contributing to global warming ... challenging these companies at every level—from the production and marketing of fossil fuels themselves, to their underhanded political influence, to their PR prowess, to the unjust "solutions" they propose, to the fossil fuel based globalization they are driving."[38] The movement works through popular education and protest, and seeks to provide a space for the articulation of claims by those most affected by climate change, but who contribute least to the problem. Among the Climate Justice activists with a strong presence at the 2002 UN COP in Delhi were fisher folk, farmers, and others whose livelihoods are being affected by climate change. The protests raised critical issues of whose voices were being heard and whose interests were being served by the sort of marketized governance solutions being proposed in the formal negotiating arenas.[39]

Governance issues and challenges

From this discussion of the politics of responsibility, costs, and burdens of climate change we find that there are a number of cross-cutting governance issues and challenges that arise. First, battles over the definition of the nature of the problem of climate change affect how and where it is addressed. In this chapter we have seen how action on climate change has been presented as a trade opportunity by some (and as barrier to trade by others), and how it has got caught up in, and at the same time redefines, older debates about issues of aid and development. As connections to other issues and policy arenas are made, new actors are enrolled in climate governance. The new resources being made available to tackle climate change attract international organizations, civil society groups and business alike. Carbon entrepreneurs seek a share of the market, while an increasing number of NGOs become engaged in climate politics, and international agencies from UNDP to the United Nations Industrial Development Organization (UNIDO) line up alongside the World Bank to make sure their presence is felt and their expertise brought to bear.

Just as each of these actors seeks to define the nature of the problem in a way which advances their interests, so too they compete to define appropriate solutions. In particular, the privileging of market-based solutions reflects the interests of certain organizations and nation-states, with their neo-liberal emphasis on using the market and allocating property rights to create incentives to reduce emissions wherever it is cheapest to do so. The proliferation of actors and sites for climate governance also raises significant challenges in terms of policy coordination and coherence. Instead of coordinated strategies between global bodies working in relevant areas and across levels of governance from local government up to the global, we find high levels of incoherence. The activities of one body or set of organizations can systematically undermine those of others. New trade agreements increase the transportation of goods over longer distances, adding to the emissions that climate negotiators are struggling to reduce,[40] while multilateral development bank-lending supports projects that commit vast amounts of GHG into the atmosphere. For instance, the World Bank-supported, $4.14 billion coal-powered "Ultra Mega" 4,000 megawatt power plant in Gujarat, India, will emit more carbon dioxide annually than the nation of Tunisia, according to the US Department of Energy.[41] We have seen in this chapter that whereas once climate change was an issue mainly addressed by the UN, its nature as a political issue at the intersection of the relationship between economy-energy and environment means that it touches many other policy areas which in turn shape the effectiveness of action on climate change. This close interweaving of climate change with other key sectors of the economy raises another important issue, that of the "ungovernance" of climate change: issues and areas of active and deliberate neglect. These are areas which are often not considered as climate policy, but which contribute significantly to the problem. This requires us to look not only at the activities of governments, but at the international institutions and market actors that shape so strongly the direction of national development strategies.

Second, this chapter has shown how new sites for climate governance become venues for the articulation of demands that it has not been possible to achieve by other means or in other fora. The forging of links to aid, debt, and trade issues illustrates this clearly. While some have sought to maintain clear boundaries around climate governance for fear of overloading the agenda with the politics of these other issues and hence slowing progress, these new sites for climate governance also open up new potential avenues for advancing action. While the interlinkages between different policy issues create a challenge in terms of the governability of climate change, positive benefits might be achieved from such synergies. The emergence of new sites of climate politics is not, however, just about constructing new avenues of cooperation. Older development concerns re-emerge in the climate context. Concerns about subsidies to failing technology sectors whose applications are inappropriate to local Southern contexts for which they were not designed; the use of conditionalities and tied aid which requires recipients to purchase the technology from the donor, have all resurfaced as concerns expressed by developing countries in the climate change debate.[42]

Third, we find that issues of process have been decisive in shaping climate governance outcomes. Who gets to define issues of responsibility, equity, and justice is a function of who participates and who is represented. Many smaller developing countries have limited representation in the international negotiations, while many more powerful developing countries also operate in other fora by engaging with, for example the APP, where they expect higher returns from their engagement. As climate governance enrolls more actors in complex and disparate ways, questions of accountability come to the fore—whether it is the accountability of CDM project developers to communities expected to host projects, the accountability of funders such as the World Bank and GEF to countries and communities, or compensation for vulnerable areas of the world as accountability for climate impacts. This creates enormous challenges for the design of institutions that are transparent and open to participation, yet at the same time robust and effective. Indeed, the need to enhance the effectiveness of action across scales and improve the accountability of key actors in climate governance has provided the point of departure for many of the transnational networks that are the subject of the next chapter.

3 Between global and local
Governing climate change transnationally

As we outline in the Introduction, climate change is an issue of concern not only on international and national agendas (Chapters 1 and 2), but also for an array of transnational networks.[1] A growing number of arrangements which cross national boundaries are being formed by a range of actors with the specific purpose of addressing climate change. In this chapter we explore this transnational phenomenon. First, we examine the roots of transnational climate change governance and the governance "functions" that such networks perform. Second, we consider different types of transnational climate change governance networks—those based purely on the involvement of public actors, those that involve a hybrid of public and private actors, and those that involve only private actors. In the final section, we consider the issues and challenges for governing climate change that this phenomenon raises.

Transnational networks and climate change

Transnational relations are not new. In the early 1970s Keohane and Nye[2] were among the first to draw attention to the significance of these cross-border transactions and networks, which seemingly bypassed the central organization and control of national governments. As the debate on global governance (see Introduction) gathered pace during the 1990s, a range of scholars pointed to the importance of transnational coalitions and networks in influencing the development of policy in various spheres of global affairs and in undertaking governance in their own right.[3] The presence of arrangements that sought to govern global environmental issues outside of the remit of international regimes, and without the necessary involvement of nation-states, provided one of the foundations for the argument that a different way of approaching global politics was required.

The emergence of transnational climate change governance

In the climate change arena, the presence of such networks can be traced back to the beginning of social and political concern for the issue. For example, transnational networks of municipal governments such as the Climate Alliance and ICLEI's[4] Cities for Climate Protection (CCP) originated in the early 1990s as both climate change and sustainability began to become significant issues for local authorities. However, most transnational arrangements are more recent, postdating the 1997 Kyoto Protocol, and in the ensuing decade their numbers and profile have increased significantly. Why might this be the case? Karin Backstränd[5] suggests that the recent growth of transnational arrangements can be seen as a "response to the regulatory deficit or implementation deficit permeating multilateral regimes." In other words, the emergence of transnational governance could be regarded as a response to the perceived failures and limitations of more traditional international institutions and national governments in addressing the problem of climate change, some of which we have discussed in Chapters 1 and 2. Taking a slightly different line of argument, Andonova et al.[6] point to the specific characteristics of the climate change issue as providing fertile ground for the emergence of transnational governance networks. First, the diversity of national commitments in the international climate regime creates opportunities for alternative forms of cooperation between actors subject to international targets and those who are not. Second, the inclusion of flexible mechanisms in the Kyoto Protocol has led to the involvement of a range of new actors working across borders in the design and implementation of international climate policy. We saw in Chapters 1 and 2 how, despite being market mechanisms, they still involve high levels of state involvement and require a range of governance functions to be performed by project developers, mediators and verifiers. Third, the complexity and breadth of climate change as a policy agenda offers potential for issue- and sector-specific collaborations alongside international agreements. In short, the nature of the climate change governance problem and how it is being addressed has created the political space for new collaborations and mechanisms of governance, such as the transnational initiatives that are our focus in this chapter.

However, it is important to realize that the emergence of transnational climate change governance has not taken place in a political, social, or economic vacuum. As we have seen in Chapters 1 and 2, broader trends—frequently referred to by terms such as globalization and neo-liberalism—have both provided the means and the rationale for the involvement of a range of actors in seeking to shape the governance of a range of local to global problems. Across many nation-states, principles of neo-liberalism have been used as the basis for reducing the role of the state and increasing the role of the private sector in service delivery. Various forms of partnership have been promoted as the means through which to determine public policy goals and deliver governance. In the environmental arena, the argument that responsibilities for governing global issues should be shared between public and private actors was epitomized by the 2002 World Summit on Sustainable Development (WSSD) which established the role of so-called Type II partnerships (between public and private actors) in delivering sustainability.[7] This is a role that many non-state actors have actively sought, on the basis of arguments of their relative expertise, ability to engage with a range of different communities, and for reasons of corporate social responsibility (see Chapters 4 and 5). We suggest that it is this context, together with the specific characteristics of climate change as a governance problem, which has led to the emergence of transnational climate change governance.

Governing transnationally

Having recognized that transnational networks now form an important part of the governance landscape for climate change, the critical questions that emerge are whether, how and why such networks are able to have any effect. While international regimes and national governments can deploy binding agreements and various policy instruments, such as economic incentives or sanctions, transnational networks do not at face value seem to have much at their disposal in terms of the standard means of shaping behavior or ensuring compliance with particular targets, goals, and norms. However, we can identify a range of "soft" mechanisms and instruments that networks deploy, through information-sharing, capacity-building and implementation, and rule-setting.[8] These three governance functions are not mutually exclusive—some transnational networks may use all three approaches, while others may focus on one particular function (see Box 3.1).

   Within transnational networks, information-sharing can perform crucial governance functions by framing issues, setting agendas, defining what counts as responsible or effective action, offering inspiration, and providing a means of benchmarking achievements. In the context of transnational governance, information-sharing is usually orientated towards network constituents and acts both to reinforce engagement and to police its boundaries. A second set of governance functions

Box 3.1 Examples of transnational governance

Information-sharing

The Climate Group's statement of principles declares that members will: "share lessons learned with others and keep apprised of best practices, policies and new innovations." (*The Climate Group Principles*, www.theclimategroup.org/assets/The%20Climate%20Group %20Principles.pdf; accessed October 2009).

Implementation and capacity-building

The Renewable Energy and Energy Efficiency Partnership (REEEP) "selects projects which can be replicated and scaled-up, and have a real impact on the development of the market for renewable and efficient energy and innovation. The partnership has funded 130 projects in more than 65 countries, primarily in the emerging markets and the developing world." (*REEEP Activities*, www.reeep.org/ 46/reeep-activities.htm; accessed October 2009).

Regulation

The Regional Greenhouse Gas Initiative (RGGI) has members from 10 states of the USA in the north-east and mid-Atlantic regions and is seeking to create a regional cap-and-trade program for CO_2 emissions from power plants among its members. Participants have developed a Model Rule to guide each state in the development of individual regulatory and legislative proposals necessary for implementation (*About RGGI*, www.rggi.org/about; accessed October 2009).

relating to implementation and capacity building is regarded as particularly important in those transnational networks which have emerged as a result of the 2002 WSSD and other multilateral environmental agreements.[9] Through the provision of resources, transnational networks seek to develop the capacity of their constituents to act on climate change by providing access to expertise, funding, or technologies that would otherwise be unavailable. In some cases, transnational networks are seeking to develop the capacity to implement or respond to international or national policy frameworks, such as contributing to meeting targets set in the Kyoto Protocol or being able to participate in the CDM. In other cases, transnational arrangements are seeking to build capacity in response to network aims and goals. A third set of governance functions that transnational networks undertake involves regulation. This can take place through setting particular standards, such as goals for emissions reductions or certification schemes, establishing a particular basis for membership (for example, signing up to a specific program of actions or a code of conduct), or various means of measuring performance. Networks vary considerably in the extent to which they use and implement regulations. However, despite the supposedly voluntary nature of transnational climate change governance, the presence of regulation and (sometimes) sanctions suggests that networks are able to exert a degree of power over their members.

In the absence of the traditional governance toolkits of nation-states and international institutions, transnational governance arrangements have devised a range of means through which they seek to guide and direct their constituents towards particular goals. At one end of the spectrum, practices of sharing information could be regarded as entirely voluntary, yet in the process of establishing what does and does not count as best practice, for example, transnational arrangements can exert considerable pressure on members to conform to particular ways of working or to certain norms. Once one or two key market players participate in a scheme like the Carbon Disclosure Project (CDP) (see Chapter 4), there is peer pressure on others to show investors that they are willing to report on their emissions profile. Sharing information in this manner can serve to define the agenda, and determine how the climate change problem should be tackled. Means of building capacity could equally be regarded as voluntary—members are free to join in with project proposals, to take or leave advice, or to disregard funding opportunities. Again, however, by being able to determine the criteria upon which funding is distributed, to set the rules of the game, and by providing certain forms of advice or access to particular sorts of technologies or information, networks can also find themselves in powerful positions with respect to their constituents. Transnational networks also undertake explicit forms of regulation, establishing standards and disclosing performance, developing membership rules and, on occasion, sanctioning those who fail to comply. Despite their apparently voluntary nature and weak position, through these means transnational networks are able to govern by shaping the ideas and behavior of their constituents in accordance with certain norms and goals for addressing climate change, suggesting that they warrant further exploration as part of the contemporary landscape of climate governance.

Comparing types of transnational climate change governance

Transnational climate change governance arrangements involve a whole host of actors—from municipal governments to private foundations, non-governmental organizations to multinational corporations. These networks take on different characteristics depending on whether they are constituted by public actors (e.g. sub-national governments), private actors (e.g. corporations), or a hybrid of public and private actors. Here, we outline the different types of transnational network involved in governing climate change and compare their roles with respect to the functions of information-sharing, implementation and capacity-building, and regulation discussed above.

Governing beyond but through the state: public transnational governance

Transnational environmental governance may involve networks of public actors including, for example, sub-units of central government, regional or state authorities, and city or local governments. In the climate domain, examples include transnational municipal networks, discussed in more detail below, networks of regional governments, such as the New England Governors and the Eastern Canadian Premiers (NEG-ECP) 2001 initiative to adopt a joint Climate Change Action Plan[10] and the Network of Regional Governments for Sustainable Development (nrg4SD), as well as bilateral agreements between sub-national governments to act on climate change, such as the initiative between the states of California (USA) and Victoria (Australia).

Municipal networks provide one of the first and most extensive examples of transnational governance. In the early 1990s three transnational municipal networks—Climate Alliance, ICLEI CCP, and Energie-Cités were established with member cities in Europe and the United States.[11] Founded in 1990 as an alliance between European cities and indigenous peoples, Climate Alliance has some 1,100 members in 17 European countries with a concentration in Germany, Austria and the Netherlands. Its aim is to reduce emissions to 50 percent below 1990 levels by 2030 and protect rainforest through partnerships and projects with indigenous rainforest peoples.[12] CCP was formed in 1992 as an initiative of ICLEI.[13] Municipal membership of the CCP network initially reflected its origins in North America and Europe, but has since expanded with specific campaigns in Australia, Canada, Europe, Latin America, Mexico, New Zealand, South Africa, South Asia, Southeast Asia, and the United States, and over 1,000 members worldwide,[14] accounting for "approximately 15 percent of global anthropogenic greenhouse gas emissions."[15] CCP members pledge to reduce their emissions of GHG by between 10–20 percent from 1990 levels by 2010. Energie-Cités stemmed from a project funded by the Commission of the European Union and is a somewhat different network, with an explicit focus on local energy policy in which addressing climate change is but one factor. Founded in 1990 and based in France, it now has over 160 individual members in 25 European countries, with a concentration in francophone countries. While its members are primarily municipalities, the network also includes local energy management agencies, municipal companies, and groups of municipalities. In this way, its reach may be greater than its membership numbers suggest.[16]

In the mid-2000s, just as membership and interest in these initial networks appeared to be stagnating, a new wave of transnational municipal networks emerged. Following the success of the 2005 US Mayors' Climate Protection Agreement,[17] in which cities pledge to meet or beat the Kyoto Protocol targets in their own communities, ICLEI has supported the development of the World Mayors' Council on Climate Change with the explicit purpose "to politically promote climate protection policies at the local level."[18] The emergence of the C40 Climate Leadership Group (C40) also signals the more avowedly political nature of this second wave of transnational municipal networks. This network was instigated by the mayor of London and the Climate Group (TCG) and formed by 18 cities in 2005 as a parallel initiative to the Group of Eight (G8) Gleneagles summit on climate change. In 2007 this network entered into a partnership with the Clinton Climate Initiative (CCI) and expanded its membership to include 40 of the largest cities in the world.[19] Activities which this network is undertaking include collaboration with Microsoft to produce software for greenhouse gas emissions accounting at the city scale, and the Energy Efficiency Building Retrofit Programme, which "brings together cities, building owners, banks, and energy-service companies to make changes to existing buildings to reduce greenhouse gas emissions."[20] While ostensibly a network of municipal governments, the central involvement of both the Clinton Foundation and TCG indicate that, in practice, the boundaries between public and private actors in transnational climate governance are increasingly blurred. These new developments have helped to raise the profile of municipal responses to climate change internationally. At COP 13 in Bali, municipalities formed the second largest delegation and signed the Bali World Mayors and Local Governments Climate Protection Agreement to tackle both mitigation and adaptation.

Across these various transnational municipal networks, it is possible to find evidence of information-sharing, implementation and capacity-building, as well as rule-setting. Information-sharing functions "are the bread and butter of [transnational municipal networks]. Networks are frequently established for the explicit purpose of creating and sharing "best" or "good" practice, and municipalities indicate that the opportunity to learn about "what works" from other places is a key motivation for their participation."[21] For example, transnational municipal networks frequently collate examples of best practice in databases, disseminate case studies through newsletters, or may even, as is the case with Energie-Cités, organize study tours where groups of local officials and politicians seek to learn at first hand how local energy policies were created.[22] Municipal networks also undertake various forms of audit and inventories of municipal greenhouse gas emissions and action plans which are made available to other members. Such strategies have the advantage of being open and inclusive, as well as avoiding the need for explicit forms of intervention or the use of more hierarchical forms of authority within the network. However, their success as governance strategies is open to debate. There is little evidence for example, that the recognition and dissemination of best practice leads to action in a direct sense,[23] or that improved knowledge of local emissions necessarily leads to policy implementation.[24] Rather, these sorts of information-sharing activities are often most effective in terms of framing a policy problem, in order to persuade others about the need for, and possibilities of, action. By framing the nature of the climate policy problem and its appropriate solutions, strategies of information-sharing both serve to delimit the field of action and offer a means of monitoring performance among network members,[25] in turn making the problem of climate change governable at the municipal level.

Transnational municipal networks are also involved in governance through implementation and capacity-building. Given that municipalities are not directly charged with implementing international environmental agreements, such as Kyoto Protocol, such activities are primarily orientated towards achieving national or local policy objectives as well as meeting the goals established by the network. In the European context, one mechanism through which this strategy has been pursued is through securing funding for particular projects. For example, the Climate Alliance is funded for the coordination and evaluation of the European Mobility Week.[26] Securing additional, external funding is also important for other municipal networks. One of the key features of the C40 network has been the partnership with the CCI and leverage of funding for a program of energy efficiency improvements in office buildings. Transnational municipal networks are also able to access national funding, either to set funds aside specifically for the program, such as is the case with CCP Australia,[27] or by working with members to access existing funding programs.

The regulative functions of transnational municipal networks are particularly interesting, given the public actors involved in these networks and the traditional association between regulation and forms of public authority. Several different forms of standard- and rule-setting are apparent. The first relates to forms of recognition, in which meeting particular standards of performance are explicitly rewarded by the network. One such example is the "Climate Star" awarded annually by the Climate Alliance to municipalities for their achievements.[28] A second regulative strategy employed by transnational networks has been that of benchmarking. The CCP network is based around a series of five milestones of progress, involving conducting an emissions inventory, setting an emissions reduction target, formulating an action plan, implementing policies, and monitoring progress.[29] Rather than being based on hard rules, the regulative function of climate governance within municipal networks is more akin to processes of voluntary standard-setting, where compliance is achieved both as a result of the recognition on the part of the municipality of the authority of the network and through processes of peer pressure and competition.

Towards partnership? Hybrid transnational governance

A second type of transnational climate change governance we can term hybrid, involving actors from the public and private spheres in various forms of collaboration. Central to the emergence of hybrid transnational governance has been the growing emphasis in multilateral fora, international organizations and national governments, as well as among some sections of the business and NGO communities, on public-private partnerships (PPP) as a means of effecting global governance.[30] The WSSD held in Johannesburg in 2002 promoted the use of these so-called Type II agreements as a means of pursuing sustainable development alongside formal international (Type I) agreements. Bäckstränd[31] defines PPP as "institutionalized cooperative relationships between public actors (governments and intergovernmental organizations) and private actors (corporate and civil society actors)." These formalized PPP are an important feature of the transnational climate change governance landscape. However, in the context of growing neo-liberal orthodoxies about the role of the state and other actors, more informal hybrid arrangements have also emerged through the collaboration of public and private actors as a means of governing climate change.

Turning first to formalized PPP, of the 334 Johannesburg Type II agreement authors suggest that approximately 90 are concerned with issues of energy and climate change, though perhaps only 34 are explicitly directed to the issue of climate change.[32] One of the largest and most complex is REEEP. Formed on the initiative of the UK government in the immediate aftermath of the WSSD, REEEP is organized through an international office and eight regional secretariats. Its membership comprises national and sub-national governments, international organizations, businesses, and NGOs. Its aim is to "to accelerate the global market for sustainable energy by acting as an international and regional enabler, multiplier, and catalyst to change and develop sustainable energy systems."[33] In terms of its governance functions, REEEP engages both with information-sharing and implementation and capacity-building. The regionalized structure of REEEP is intended to capture existing knowledge about the barriers and opportunities for enhancing renewable energy and energy efficiency, and secretariats undertake various dissemination activities and events. At the same time, drawing on funding from various national governments, REEEP is involved in the selection and support of projects to develop new technologies and markets for renewables and energy efficiency. While such projects may assist in helping countries to meet their commitments under international frameworks for climate change (i.e. the UNFCCC and Kyoto Protocol), the focus here is more on developing capacity in the energy sector more broadly, especially with regard to changing the regulatory and market conditions for the implementation of energy efficiency and renewable energy technologies.

Another arena within which formal PPP have been created is in the implementation of one of the Kyoto flexible mechanisms, the CDM (see Chapter 1). While the formation and regulation of the CDM process takes place within the international arena, Bäckstrand[34] suggests that CDM projects, which now number more than 1,500, can be regarded as public-private partnerships because of the range of actors, including host governments, private investors, carbon brokers, and NGOs involved in their design, implementation and verification. Streck describes the CDM as "an innovative model of cooperation between the private and public sectors."[35] The collaborative network structure in which state and non-state actors collaborate in a partnership arrangement confers on non-state actors, such as the approved auditing bodies known as Designated Operational Entities, "a variety of voluntary, semi-formal and formal roles in formulating policy responses and implementing international agreements."[36] However, as we discuss in more detail in Chapters 4 and 5, the ways in which such forms of governance operate in practice is highly disputed, demonstrating that the function of implementation cannot be understood just as a means through which internationally agreed goals are transmitted to particular places, but rather is a means through which what constitutes just and appropriate responses to climate change are constructed and contested. In this light each of the actors involved in the CDM project cycle has a significant role in determining what counts as a reduction in GHG and the extent to which sustainable development benefits are realized. Because it operates as a market mechanism, credits in the form of Certified Emissions Reductions (CERs) are issued for reductions in GHG that would not have otherwise been possible, but no monetary value is attached to sustainable development benefits. This explains why investors have prioritized projects which earn them the most CERs without necessarily delivering the greatest benefits to communities that host the projects.[37] The CDM and other initiatives, such as projects implemented under the auspices of the APP, highlight the intertwined nature of the international regime and transnational governance. While ostensibly created by and for nation-states, both the Kyoto Protocol and the APP are reliant on the emergence of transnational hybrid networks for their implementation.

As well as being driven through multilateral processes, hybrid transnational governance has also emerged from the bottom up. One such example is TCG, a UK-based network. Launched in 2004 by then prime minister Tony Blair, TCG was initiated by the Rockefeller Brothers Foundation as a means of fostering dialogue between those corporations and sub-national governments in the United States who were taking a progressive approach to addressing climate change and their European counterparts. Membership is on a fee-paying basis and now includes over 50 members from businesses, state, and regional governments, organized through a UK headquarters and regional offices in the USA, Australia, India, and China. TCG has undertaken a range of information-sharing activities, including the production and promotion of examples of best practice, publicity events about the significance of climate change and forms of peer-to-peer learning. In addition, the network has been involved with various projects aimed at implementation and capacity-building, including the States and Regions Initiative, which seeks to "enable governments to overcome obstacles in the implementation of climate actions"[38] as well as regulative functions through the development and promotion of the Voluntary Carbon Standard (VCS) (see Chapter 4).[39] The complexity and range of the different forms of governance taking place within TCG, as well as formal partnerships such as REEEP and the implementation of international policy instruments such as the CDM, suggest that such initiatives go beyond the sort of organizational structures of non-governmental and business coalitions that characterized the landscape of climate governance in the 1990s. Instead, such initiatives are forging what John Ruggie[40] terms a "new global public domain," "an increasingly institutionalized transnational arena of discourse, contestation, and action concerning the production of global public goods, involving private as well as public actors." The emergence of this transnational domain of climate change governance raises significant issues in terms of legitimacy and effectiveness, to which we return below.

Self-regulation and private transnational governance

Alongside hybrid and public forms of transnational governance, private initiatives and networks are also contributing to the emergence of the "new global public domain."[41] Scholars have documented private transnational governance in a number of fields including forestry, water, chemicals, and food.[42] Private transnational governance involves a variety of non-state actors, including corporate and civil society sectors, in arrangements through which "issues are defined, rules are made, and compliance with these rules is monitored"[43] beyond the direct purview of nation-states or other actors from the public sphere.

One means by which private transnational governance has emerged with respect to climate change is through the work of business leadership groups, orchestrated by bodies such as the World Business Council on Sustainable Development (WBCSD) or "think tanks" such as the World Resources Institute (WRI). One initiative established by the WBCSD and led by the Swedish energy company Vattenfall is Combat Climate Change (3C).[44] This initiative aims to establish a global opinion group on the need for international action on climate change, but in so doing asks its members to commit to a series of principles about the nature of the climate change problem and how it should be governed, in effect leading to the development of norms concerning climate change governance in the private transnational arena. A second means through which private transnational governance networks are being forged is through partnership working between corporate and civil society actors. The Climate Community and Biodiversity Alliance (CCBA) was founded in 2003 by Conservation International, and includes corporations such as BP and Intel as well as other NGOs, such as Rainforest Alliance and the Nature Conservancy Council.[45] The CCBA has established the CCB standard for land management projects that simultaneously address climate change, social goals and biodiversity conservation. Since the standards were released in 2005, six projects have been validated, 24 are in process, and several dozen are planning to seek validation.[46] As well as performing a regulative function through the promotion of voluntary standards for land management, the CCB also serves an implementation and capacity-building function through enabling action on elements of international climate policy, including the CDM and Reducing Emissions from Deforestation and Forest Degradation (REDD). The promotion of standards for carbon offset markets has been another key area in which private governance initiatives have flourished, which we discuss further in Chapter 5.

Alongside the growth in hybrid forms of transnational governance, it is perhaps the growing involvement of private actors in governing what many perceive as a fundamentally public issue—the global climate—that has raised the most concerns. On the one hand, critics have argued that the increasing involvement of non-state actors and, in particular, private sector organizations, in climate governance is detrimental in terms of the legitimacy of, and accountability for, decision-making. This is seen as part of a broader trend towards the privatization of governance in which private actors come to dominate public governance institutions and assume more "government-like" functions themselves.[47] On the other hand, commentators have also pointed to the highly fragmented and limited nature of transnational governance arrangements—despite the growing numbers of municipalities, regions, businesses, and NGOs involved, the vast majority remain outside of this "new global public domain."[48] While transnational climate change governance may be interesting in principle, to date its capacity to make a difference in terms of reducing emissions of greenhouse gases may seem limited.

Governance issues and challenges

Transnational initiatives for addressing climate change are now firmly established on the governance landscape. Their presence signifies the broad and growing constituency for whom climate change is an issue of importance, and may be read as opening up the climate change domain to a wider variety of actors and to different perspectives on addressing the problem. However, the emergence of transnational climate change governance also raises a number of challenges.

First, we can see from this chapter that in terms of the impact and effectiveness of transnational climate governance, the evidence is mixed.

Given the short timescales over which they have been operating, the lack of common baselines and the multiple measures used to record reductions of greenhouse gas emissions, assessing the difference that such initiatives are collectively making in terms of addressing climate change is almost impossible. However, there is some evidence from particular initiatives that impacts are being achieved. For example, ICLEI Australia suggest that the CCP program resulted in a saving of 4.7 million tonnes of carbon dioxide equivalent in 2007/8, and a total of 18 million tonnes since the program started in Australia in 1998/99.[49] Two other features of transnational initiatives also suggest that they may have an important impact on governing climate change. First, many networks and partnerships include targets and timetables for reducing emissions of greenhouse gases that go far beyond those agreed internationally. This may create momentum for action towards these ambitious goals, and ease the path of any future international agreements. Second, many such initiatives target emissions of greenhouse gases that are not covered by international and national climate policy, which has tended to focus on emissions reductions from the energy and industrial sectors. For example, initiatives such as C40 focus on energy consumption in the commercial building sector, while others concerned with regulating carbon offsets address the climate impacts of individual consumption. Nonetheless, a key challenge here is the additionality of these initiatives—are they mobilizing action that would otherwise not have taken place, or acting as window dressing for actions already underway? Either way, there is certainly no evidence that such initiatives are any worse than international agreements in terms of their impacts and effectiveness, and may, by achieving modest emissions reductions, be creating "coalitions of the willing," and raising expectations about what it is possible to achieve, be acting in a way that is complementary to the international regime. If however, such initiatives were to act as a distraction from international agreements, or a substitute for them, the outcomes in terms of the effect on addressing climate change might be much less benign. It is exactly this fear which underpins concern about the APP diverting the attention of key actors away from the need for a legally binding emissions reductions agreement to replace Kyoto.

We have also seen that transnational climate governance initiatives also raise significant challenges in terms of accountability. Once the governing of climate change moves from discernible public arenas of government, questions concerning who is participating, on whose behalf and how they are held to account are inevitably raised. Conceiving of accountability as involving both answerability—the need to justify and explain positions and actions adopted—and enforceability—the ability to impose sanctions for non-compliance—can help to examine the complex questions raised by transnational climate governance.[50] On the one hand, such initiatives may have done much to promote the answerability of a range of actors in terms of their actions on climate change. One feature of many of the partnerships and networks discussed in this chapter is their focus on making their actions accountable through mechanisms such as audit, disclosure, peer review, standard-setting, and other forms of soft regulation. In this way, they may serve to increase the accountability of various actors to new constituencies concerned with climate change. On the other hand, it is clear that the ability of such initiatives to enforce standards and to implement particular actions is limited. In the absence of forms of sanction, network members or constituencies may have little recourse should action not be forthcoming. In this manner, transnational climate governance has much in common with the international regime, which has both raised the issue of the answerability of nation-states in addressing climate change while also demonstrating its weakness in enforcing the targets and timetables agreed.

One of the consequences of the partial accountability of transnational climate governance is that, unlike the international negotiating fora and national policy contexts discussed in Chapter 2, issues of equity are rarely visited. Issues of procedural equity—who is able to participate in such initiatives—and distributive equity—what the impacts of such initiatives might be in terms of social and environmental justice—are critical. Turning to the equity of participation, it is clear that initiatives vary significantly in terms of their inclusivity and the sharing of the benefits of network participation. Some, such as CCP and Climate Alliance for example, are ostensibly open, seeking to recruit as many members as possible. Others are based on an exclusivity of membership, for either geographic or strategic reasons. However, even in the case of open initiatives, there are significant barriers to involvement, including the capacity required to set up membership and access funding, as well as the way in which the problem of climate change governance has been framed, which has to date mainly been in accordance with Northern concerns for mitigation and energy efficiency, rather than, say, issues of adaptation and the provision of basic services, though this picture is now changing. In terms of the equity of impact, there has to date been little analysis of the implications of transnational initiatives in terms of issues of social and environmental justice. As we discuss in more depth in Chapters 1, 2 and 4, CDM projects have been found to have mixed benefits for the communities in which they are based, and it is likely that projects carried out under the auspices of other forms of transnational governance may have similar impacts. At the same time, given the partial nature of involvement in such initiatives, attempts to govern climate change in one location or firm may serve to push "dirty" industries outside these jurisdictions, with negative consequences for other communities without the power to resist them. Such impacts are, of course, not unique to transnational forms of climate governance, but do suggest that their wider implications for sustainable development and socio-environmental justice need to be considered in their development and implementation.

Overall, we find that transnational climate change governance raises significant issues in terms of effectiveness, accountability, and equity. In part, these challenges arise because such initiatives transcend and cut across the frameworks through which we have traditionally analyzed the politics of global affairs. Consequently, there are three ways in which the phenomenon of transnational governance requires us to reconsider the notion of global governance. First, in relation to the spatial organization of global politics. Rather than taking place within discrete, and hierarchically arranged, levels of governance, the transnational governance phenomenon is complex, where so-called local initiatives may have a reach across national jurisdictions, and international agreements give rise to new forms of regional or local partnerships. This finding reinforces our conclusions in Chapter 1 that tracing the location of climate governance is no longer a straightforward issue. Second, with respect to the relationship between state and non-state actors, transnational governance initiatives raise questions as to whether these demarcations are analytically useful, given their hybridity, the mixing of roles, and the extension of forms of public authority into ostensibly private domains and vice versa. As we saw in Chapter 2, the increasing crossover between policy issues, between institutions and the proliferation of actors involved in climate governance calls into questions where the boundaries between the public and the private might lie. Finally, transnational governance initiatives demonstrate not only the importance of the power of one set of actors over another, as currently captured in many analyses of international relations, but also in a more dynamic sense, where power is exercised in the framing of the climate problem, and in the day-to-day activities of networks and partnerships where power sharing and agreed norms of conduct are key to their success. Conceiving of climate governance as taking place outside of the international regime in transnational initiatives, therefore, raises some fundamental questions about how, where and why the governing of global affairs takes place, issues to which we return in Chapter 6.

4   Community and the governing of climate change

One long-standing feature of environmental governance in general is the involvement of community-based and "civil society" organizations in all aspects of politics as advocates, educators, implementers, and active citizens; as makers as well as shapers of policy.[1] It is unsurprising, then, that a wide array of social or community actors, including various voluntary groups, NGOs, and civil society institutions, have also been critical to the politics of climate change.[2] Because addressing climate change implies action at all levels, and, as we show in the Introduction, change in everyday patterns of production and consumption, engaging communities in mitigation and adaptation efforts, is also a key policy goal. At the most basic level, without the participation of the public—for example in terms of their acquiescence to new forms of carbon regulation, their participation in carbon markets, or changing behavioral practices at home and at work—climate policy will fail to achieve the targeted reductions in GHG emissions. Urging communities to "do their bit," national and local governments, as well as a range of non-state actors, have sought to engage communities in responding to climate change. Social responses to climate change— much like those of the market (Chapter 5)—therefore involve a complex mixture of public and private authority, top-down direction, and bottom-up experimentation.

In this chapter we focus on this spectrum of community-based responses and their role in the governance of climate change. First, we consider the ways in which governments have sought to engage the public in climate change. Here, we document a shift from a concern to educate individuals in order to change behavior towards various attempts to draw community expertise into the design and development of climate change responses. Second, we examine the ways in which place-based communities are being mobilized in response to climate change by (government) authorities. All too often it seems, such schemes fail to engage with communities and tensions emerge between the realization of global objectives—e.g. the reduction of GHG in the atmosphere—and locally based concerns and needs. Third, we turn to the example of Transition Towns to assess the emergence of grassroots responses to climate change. While concerns for social and environmental justice lie closer to the heart of such initiatives, we find that tensions still remain in seeking to govern the global and the local simultaneously.

Engaging communities: from ignorance to expertise

Since it first emerged on the policy agenda in the early 1990s, governments (and some non-governmental actors) have sought to mobilize community participation in the climate change issue. Despite some apparent consensus on the importance of engaging the public,[3] there are different views about what this should entail. Traditionally, it was thought that behavioral change would occur if communities were provided with information about the nature of environmental risks and what they could do to address them. More recently, this approach has been critiqued and the importance of basing policy on community knowledge and concerns has been recognized.[4] In this section, we trace this shift and its implications for governing climate change.

As public concern for environmental issues grew during the 1970s and 1980s, an apparent paradox emerged. On the one hand, opinion polls and surveys seemed to indicate that environmental matters—and increasingly those of a global nature, such as climate change—were attracting high levels of concern among the public. On the other hand, the public appeared to be relatively ignorant about the scientific basis and implications of these issues, and at the same time unwilling to engage in behavioral change to address them.[5] In this context, the conviction emerged in policy circles that there was a "deficit" of public knowledge about the issue, spurring the development of environmental education campaigns in Europe, North America, and Australia. One example is the Helping the Earth Begins at Home campaign, run in the UK during the 1990s. In this campaign, newspaper and television advertisements provided information about the causes of global warming (as it was then commonly referred to), and about the potential for behavioral changes at home to save energy, save money and reduce emissions of greenhouse gases.[6]

Yet, despite the high levels of exposure in the media, Steve Hinchliffe[7] found that the campaign was largely ineffective. In focusing on the provision of information, the campaign, like many other initiatives, had neglected the numerous other factors which shape behavioral change, including issues of responsibility, trust, and efficacy. For example, in their exploration of public understanding of global environmental issues in the Surselva region of Switzerland, Carlo Jaeger et al.[8] conclude that knowledge about climate change is a less important factor in determining behavior than social networks and rules that sanction and enable mitigating actions. Analysis by Willet Kempton et al.[9] found that public understandings of climate change in the United States were based on fundamental moral and religious views on the relation between nature and humanity, the rights of other species, humanity's right to change or manage nature, and our society's responsibility for future generations. More recent studies have sustained this view that public engagement in environmental issues, and climate change in particular, is shaped by a range of individual and social factors, and that the provision of information alone is unlikely to have much effect on behavior (Box 4.1).

Recognition of the flaws of the "information deficit" approach led commentators to call for new approaches for involving communities that recognized and valued their experiences and knowledge, and which

---

Box 4.1 Barriers to engaging the public

Individual barriers

Lack of knowledge; uncertainty and skepticism; distrust in information sources; externalizing responsibility and blame; reliance on technology; climate change perceived as a distant threat; importance of other priorities; reluctance to change lifestyles; fatalism; helplessness.

Social barriers

Lack of action by governments, business and industry; "free rider effect"; pressure of social norms and expectations; lack of enabling initiatives.

Source: Adapted from Irene Lorenzoni, Sophie Nicholson-Cole, and Lorraine Whitmarsh, "Barriers Perceived to Engaging with Climate Change Among the UK Public and Their Policy Implications," *Global Environmental Change: Human and Policy Dimensions* 17, no. 3–4 (2007): 445–59.

facilitated debate about policy choices.[10] One reason underpinning this shift was the recognition that the public may have forms of knowledge which would prove valuable in understanding and responding to environmental risks. For example, Judith Petts and Catherine Brooks[11] suggest that lay people may be "local technical experts," using the case of how cyclists have questioned the validity of local air quality modeling and used their experience of city streets at the micro-scale to provide an alternative map of urban pollution.[12] Another reason was the recognition that the public's view of the practical, ethical and political aspects of environmental problems might be as important as understanding the scientific issues at hand. Importantly it was thought that this approach "might help not only to identify or implement solutions but to define, or reframe, what the problems actually are."[13]

One area of climate change governance where this more participatory approach is beginning to be adopted is in relation to adaptation.[14] In the Philippines, community-based disaster preparedness (CBDP) approaches have been used by the Red Cross as a means of reducing vulnerability and enhancing resilience to the impacts of climate change.[15] CBDP approaches seek to "build on existing local knowledge and experience, as well as the resources, coping and adaptive strategies of local people."[16] In the case of the Philippines, CBDP involved projects such as constructing flood defenses and training in disaster management skills in order to provide skills and experience for longer term "flexible and incremental" approaches to climate adaptation.[17] A similar approach, Vulnerability and Capacity Assessment (VCA) has been used by the Red Cross in Vietnam and Indonesia.[18] It is carried out at the level of villages and urban neighborhoods, and uses participatory rapid appraisal (PRA) tools to diagnose vulnerabilities, assess a community's risk priorities, and work together with the people to devise ways of increasing their capacities to resist hazard impacts.[19] Articulating one of the central arguments voiced in support of such participatory and deliberative techniques, Maarten K. van Aalst et al.[20] argue that one of the key purposes of the approach is to

> catalyze a process that empowers the people in the community and supports their capacity to alter their own situation. Through engagement with the grass roots, the activities that emerge will have the people's "ownership" and participation, be based on trust and therefore have more chance of success.

Their investigation of the use of such approaches found, however, a tension between the aims of local community empowerment and the need to engage with the science of climate change impacts, which required expert-led processes of knowledge dissemination.[21] Further tensions emerged in the process of community engagement in the Philippines, where the idea of communities being homogeneous within the CBDP discourse tended to subsume competing interests and priorities within and between community groups while also neglecting structural processes over which local communities have little if any control. At the same time, the rhetoric of building adaptive capacity implicitly assumes that such processes will create more "productive, participatory relations with government agencies."[22] In this manner, far from being autonomous from the central state, community-based initiatives provide a key means through which the state seeks to govern climate change, a point to which we return below.

Engaging communities on the issue of climate change has been seen as central to the implementation of policy at national and local levels, either because of a need to educate the public in order to achieve behavioral change (e.g. reducing car travel, reducing energy use), or because of the ways in which public knowledge might be able to inform policy design. We can therefore see that involving the public is a key means through which such actors seek to govern climate change.

Governing place, governing community

Public education and engagement do not exhaust the means by which policymakers have sought to govern climate change. The importance of "community" as a site of governance has been highlighted by governmentality scholars, who point to the ways in which the idea is used to mobilize particular practices and aspirations that authorities of various sorts consider desirable.[23] In particular, the rhetoric of community invokes a sense of "self-help" and "self-reliance"[24] in which communities are encouraged to take responsibility for addressing policy issues such as climate change. In this section, we examine two examples of such community-based responses to governing climate change—small-scale forestry projects[25] and renewable energy schemes[26]—and consider the implications for how we might understand by whom and with what implications climate change is governed.

Community forestry and carbon governance

Under the terms of UNFCCC and the Kyoto Protocol, projects which involve reforestation or afforestation—the conversion of land to forests or the rehabilitation of forests—and the consequent removal of CO_2

from the atmosphere (sequestration) can be included as a means of reaching national targets for emissions reductions under Activities Implemented Jointly (AIJ) and the subsequent CDM (see Chapters 1 and 2). The imperative to effectively manage and renew forest resources has long been a mainstay of conservation and development discourses.[27] Since the 1970s the argument has evolved that communities need to be involved in the management of forests both because of their expertise and because without such involvement efforts will be ineffective at best or deliberately undermined at worst. Attempts at "fortress conservation"— where communities are kept on the outside of conserved areas—have been shown to fail, as communities dependent upon forest resources continue to seek their livelihoods through these means.[28] The language and approach of the climate change regime also recognizes the importance of including communities in the processes of carbon sequestration, and of delivering "side-benefits" in terms of development.[29]

To date, however, only a few small-scale forestry projects have been developed under the mechanisms of the UNFCCC and the Kyoto Protocol. Some 30 forest sinks projects were tested under the UNFCCC AIJ pilot phase in which several explicitly targeted smallholders and low-income producers aimed to provide dual benefits of global climate protection and local development benefits.[30] Under the CDM a total of seven forestry projects have so far been registered, although they are more prevalent in the voluntary carbon offset market.[31] However, analysis suggests that the top-down nature of these initiatives, being driven primarily by international agencies and national governments, has meant that "pilot forest carbon projects have tended to fall short of their development objectives by failing to include local stakeholders in decision-making and to ensure their buy-in early on in the project design."[32]

The case of community forestry management in Mexico illustrates some of the complexities involved in seeking to govern climate change through the enrollment of local communities and places. In the mid-1980s, and in the context of a high rate of deforestation, the Mexican government began to promote community forestry and, more recently, has developed schemes in which local communities receive financial reward for the "ecosystem services" that the conservation and renewal of forest areas delivers.[33] In 2003, one such scheme, for Payments for Hydrological Services, was established by a government ministry and the National Institute of Ecology. Subsequently, and following the lobbying of central government by various peasant and forest-based organizations, in 2004 a program of Payments for Carbon, Biodiversity and Agroforestry Services (PSA-CABSA) was established, reflecting at least some degree of support from local communities for recognizing their forest management activities in these terms. Evidence suggests that such projects have delivered considerable economic benefits to local communities, and some notable social benefits, as well as carbon sequestration (see Box 4.2 for one example).[34] The involvement and support of communities also seems to have been relatively high, a finding explained in part by the institutional design of the scheme which involved community-based organizations in establishing the rules of the game, independent evaluations of compliance, implementation rules that have been enforced and minimized illegal logging, and rights to appeal for the providers of ecosystem services. While there are questions about the ways in which ecosystem services have been accounted for and the long-term financial viability of the PSA-CABSA, the preliminary assessment of the scheme was positive.

However, changes to the scheme in the light of international and national demands to better account for carbon services and to secure external flows of financing may serve to undermine the benefits of

---

Box 4.2 Community forestry and development

San Bartolomé Loxicha, an indigenous Zapotec community located in the state of Oaxaca, Mexico encompasses over 14,000 hectares and has over 2,250 inhabitants. The main economic activities are coffee cultivation and commercial logging. In 2004 a carbon project involving the planting of 66,000 pines in 272 hectares of the forest commons was approved for implementation. The project was considered an opportunity for complementing ongoing reforestation and conservation activities. The total investment from the program for Payments for Carbon, Biodiversity and Agroforestry Services has amounted to over Mx$2.7 million (or US$0.2 million), of which over 70 percent has reached the community. The community assembly was involved in determining how the project funding should be distributed, and 20 percent was given to the Milenio Coffee Producers Organization, most of whose 100 farmers were involved in carbon project activities, with the remaining 80 percent distributed among the farmers who participated in tree planting.

Source: Esteve Corbera, Carmen González Soberanis, and Katrina Brown, "Institutional Dimensions of Payments for Ecosystem Services: An Analysis of Mexico's Carbon Forestry Programme," *Ecological Economics* 68, no. 3 (2009): 743–61.

community involvement which have been secured. In 2006 the guidance for carbon service projects changed quite substantially, following the criteria established by the CDM concerning the size of projects involved, and the timing and extent of deforestation while funding was limited to project design with external investors required to purchase the carbon saved in these projects.[35] These changes reflected both the international guidance on forest projects from the CDM and "the interest of the Mexican government to support projects which may potentially become sources of voluntary and CDM carbon offsets."[36] Analysis suggests that this move has gone "against the interests of Mexican civil organizations," since under the new rules there "has been a reduction in the number of project applications, and an increase in the rate of applications' rejections," with the result of undermining "communities' ability to access carbon payments."[37] This example shows how decisions concerning climate governance taken at an international level can have significant implications for local communities and places, raising questions about whose resources are being governed, and whose interests are being served, by such governance mechanisms.

As we saw in Chapter 2, others have raised similar deep-seated concerns about the emerging carbon market and the ways in which communities, particularly in the global South, may be affected. Adam Bumpus and Diana Liverman[38] argue that the process of the commodification of carbon—its extraction from local contexts and circulation in global markets—is following a neo-liberal logic concerned with creating profit at the expense of both its effectiveness in terms of reducing greenhouse gases and its implications for local communities. Because of the economic logic of the CDM, and indeed of wider systems of ecosystem service payments, the implementation of CDM projects has followed the "avenues of the most profitable geographical locations," while the sustainable development benefits which are supposed to also accrue are neglected.[39] Once the value of a resource is changed and it becomes of commercial value to international investors, communities home to those resources find themselves under pressure. As Frances Seymour of the Center for International Forestry Research (CIFOR) claims, "as payments for conserving forests for carbon storage become increasingly likely, state and non-state actors alike will have strong incentives to passively ignore or actively deny the land and resource rights of indigenous, traditional and/or poor forest users in order to position themselves to claim compensation for forest stewardship in their stead."[40]

Seeking to involve communities in the governing of climate change through small-scale forestry projects has been a strategy deployed by governments and the international community, and, in part, one welcomed by some of the communities involved. However, it is a process that raises complex questions about the basis upon which communities should be involved, and about the tensions between local community needs and the imperatives of international policy and carbon markets. While on the surface a synergy between community-based forestry projects, sustainable development, and carbon sequestration appears to be a matter of common sense, the processes and practices involved reveal conflicting agendas, priorities and outcomes, and raise important questions over the nature and distribution of the benefits of such schemes.

Community power: renewable energy in the United Kingdom

In the UK, since the privatization of the energy generation and supply system in the 1980s, the development and deployment of renewable energy technologies has traditionally been assumed to be the preserve of the private sector, albeit subject to government mandate concerning the proportion of renewable energy in the supply mix. However, since 2000 and the development of the 2003 Energy White Paper, the importance of community involvement in the provision of renewable energy has increasingly been stressed. Moving beyond exhortations to consult or engage communities in decisions over the siting of renewable energy generation technologies (though this remains an important rhetoric), this new theme incorporates "notions of community led, controlled and owned renewable energy installation development"[41] and has been manifest in a series of government-funded programs, including the Community Action for Energy program, Community Renewables Initiative, Clear Skies and Community Energy scheme.[42] Research found that in 2004 there were over 500 ongoing or completed projects supported by these programs or initiatives and others with the word "community" in their title and/or rationale in the UK.[43] These included projects in which there was a degree of community ownership, through cooperatives, charities, trusts or other arrangements, or where co-ownership between a private company and community takes place. One of the best known examples of the former in the UK is the Baywind cooperative,[44] in which people within and beyond the local community become members of the cooperative and buy shares to finance the project.[45] In addition to community ownership, community involvement in renewable energy projects was manifest in two other ways: first, in terms of the involvement of communities in the development and running of projects, and second, in terms of the distribution of the benefits of such schemes, for example "providing jobs, contributing to local regeneration or providing an educational resource."[46]

Despite significant variation in how the term was used, the salience of the theme of "community" in the provision of renewable energy in the UK appears to have been high, with a number of significant funding streams dedicated to their realization and several hundred projects established or underway. In seeking to account for this phenomenon, Gordon Walker et al.[47] identify four explanations. First, developing a community basis to renewable energy generation was regarded as a means of overcoming public opposition to renewable energy, particularly in the wake of the conflicts over the siting of onshore wind farms that have occurred in the UK.[48] Second, a community-based approach allowed the government to stimulate the market for renewables, regarded as critical in the context of carbon reduction targets and the need to improve skills and capacities with regard to newer micro-generation technologies. A third rationale lay with the predominantly rural nature of the projects implemented, based on a perceived need to stimulate economic growth and security in declining rural areas. Finally, there was a sense that communities should be involved, in a normative sense, because it was a legitimate and democratic means through which decisions about energy futures should be made.[49] These explanations point once again to the significance of community as a site which is crucial for the achievement of government policy objectives for climate change. The UK government has ambitious targets for meeting 20 percent of electricity generation from renewable sources by 2020, and renewable energy generation is a key part of its climate change strategy. This case study shows how community projects become essential to, and enrolled in, these wider policy agendas, but also retain the possibility of governing climate change differently, because of the democratic mandate they engender.

Through these case studies of carbon forestry and renewable energy, we can see that the reasons for "community-based" climate governance stem not only from efforts to displace responsibilities onto the community, but also as part of a much more explicit strategy to implement policy and meet climate governance objectives devised at other scales. However, other forms of community-based climate governance—primarily emerging from outside of the state—have provided the space for alternative approaches to addressing climate change, and it is to these that we now turn.

Communities governing climate change

In addition to initiatives to engage communities that have emerged from within the state, communities have also acted as a site for independent responses to climate change. These community-based climate governance initiatives range in size and scale from individual groups of households or neighborhoods to transnational networks seeking to connect communities across national boundaries (see Chapter 3). Some are organized and orchestrated by third sector and non-governmental organizations, while others originate in the concerns and spontaneous actions of individuals. Despite their relatively low profile and limited reach, these community-based initiatives provide a means through which individuals are becoming directly involved in addressing climate change—a level of engagement that has so far eluded mainstream national and international responses to the issue. At the same time, such initiatives provide a source of alternative means of framing and acting on climate change, which offer a basis for contesting orthodox policy approaches and which may provide a source of social and technical innovation at other scales of governance. Here we consider one such example—Transition Towns—in order to examine the implications for climate governance.

Transition Towns

> A Transition Initiative is a community ... working together to look Peak Oil and Climate Change squarely in the eye and address this BIG question: for all those aspects of life that this community needs in order to sustain itself and thrive, how do we significantly increase resilience (to mitigate the effects of Peak Oil) and drastically reduce carbon emissions (to mitigate the effects of Climate Change)?[50]

The idea of a "transition" from oil-based to low-carbon societies as a localized movement originated in Kinsale, Ireland, in 2005, and was adopted by the town of Totnes, in Devon, England, in 2006 through the work of Rob Hopkins, a permaculture teacher. The central idea, as illustrated in the quote above taken from the TT website, is that communities need to face the realities of "peak oil"—preparing for a future world without oil resources—as well as significantly reducing emissions of carbon dioxide. The TT movement links these two issues together, arguing for the positive benefits that can accrue from localizing economies through, for example, the production of local sources of food and energy. Following Totnes, several other communities in southwest England took up the initiative, and in 2007 a Transition Network was established in order to support the spread of the ideas. By 2008 some 100 communities had joined the network, primarily based in the UK but also including the USA, Australia, Japan and Chile.[51] In this process, the Transition Network has become a more formal entity,[52] and a process of signing up to 16 criteria has now been adopted for those communities wishing to become part of the movement (Box 4.3). These criteria set out "demands for mainly local commitments" that "resonate with the duties inherent in communitarian theories of citizenship,"[53] in which obligations are due first and foremost to the community within which individuals reside.

A wide range of actions and activities are undertaken under the banner of Transition Town (TT), including a focus on local food production, health, transport, and energy. For example, in Totnes, a Garden Share project has been established that "matches keen, enthusiastic and committed gardeners and local garden owners who want to see their gardens being used more productively. The gardener and garden owner enter into an agreement about times, arrangements, security and the sharing of garden produce."[54] In Brixton, London, there is also a focus on food issues, with members involved in "finding out about local food suppliers and more about how to start growing their own food; reskilling (cooking, growing, permaculture, food preservation); [and] starting their own community growing and composting projects."[55] Also in Brixton, a Buildings and Energy group is working on "developing an outreach programme, building links with the London Borough of Lambeth in order to play a role in the new Local Development Framework (LDF), and contributing to the development of the Roupell Park Energy Centre,"[56] while in Berea, Kentucky, USA, the group has established a home energy conservation project.[57]

While the success of individual TT initiatives in terms of their impact on reducing emissions of GHG is both hard to measure and currently premature to evaluate, it is clear that the initiative has generated significant interest. The growth of the network, from one or two cases in 2006 to over one hundred in 2009 is one measure of its impact, but so too are the relatively large numbers of people attending information events in the Transition Towns (in many cases, several hundred people are involved in the launch or publicity events in any one town). However, whether this interest can be sustained as individual TTs move from the dissemination of information to projects on the ground remains to be seen. At the same time, the movement has provided an alternative means of framing climate change as a "governance" problem—one that focuses on consumption rather than the production of greenhouse gas emissions, and on the local rather than global scale—that has provided the basis for contesting orthodox accounts of what the climate governance problem entails. Governance experiments such as this that bubble up "from below" show that governance, even around a problem as complex as climate change, is not only constructed above

Box 4.3 The criteria for Transition Network membership

1 an understanding of peak oil and climate change as twin drivers (to be written into constitution or governing documents)
2 a group of 4–5 people willing to step into leadership roles (not just the boundless enthusiasm of a single person)
3 at least two people from the core team willing to attend an initial two day training course. Initially these will be in Totnes and over time we'll roll them out to other areas as well, including internationally. Transition Training is just UK based right now, but that's going to have to change – we're working on it.
4 a potentially strong connection to the local council
5 an initial understanding of the 12 steps to becoming a TT
6 a commitment to ask for help when needed
7 a commitment to regularly update your Transition Initiative web presence – either the wiki (collaborative workspace on the web that we'll make available to you), or your own website
8 a commitment to make periodic contributions to the Transition Towns blog (the world will be watching)
9 a commitment, once you're into the Transition, for your group to give at least two presentations to other communities (in the vicinity) that are considering embarking on this journey – a sort of "here's what we did" or "here's how it was for us" talk
10 a commitment to network with other TTs
11 a commitment to work cooperatively with neighbouring TTs
12 minimal conflicts of interests in the core team
13 a commitment to work with the Transition Network re grant applications for funding from national grant giving bodies. Your own local trusts are yours to deal with as appropriate.
14 a commitment to strive for inclusivity across your entire initiative. We're aware that we need to strengthen this point in response to concerns about extreme political groups becoming involved in transition initiatives. One way of doing this is for your core group to explicitly state their support the UN Declaration of Human Rights (General Assembly resolution 217 A (III) of 10th December 1948). You could add this to your constitution (when finalized) so that extreme political groups that have

Box continued on next page.

discrimination as a key value cannot participate in the decision-making bodies within your transition initiative. There may be more elegant ways of handling this requirement, and there's a group within the network looking at how that might be done.

15 a recognition that although your entire county or district may need to go through transition, the first place for you to start is in your local community. It may be that eventually the number of transitioning communities in your area warrant some central group to help provide local support, but this will emerge over time, rather than be imposed. (This point was inserted in response to the several instances of people rushing off to transition their entire county/region rather than their local community.) Further criteria apply to initiating/coordinating hubs – these can be discussed person to person.

16 and finally, we recommend that at least one person on the core team should have attended a permaculture design course ... it really does seem to make a difference.

Source: Criteria for becoming an "official" Transition Initiative, www.transition towns.org/TransitionNetwork/Criteria, accessed October 2009.

by policy elites where communities passively react to global political initiatives. They also proactively create alternatives and produce immediate forms of action, often fueled by a frustration with the slowness and inadequacy of existing responses.

Governance issues and challenges

In this chapter, we have suggested that alongside a growing interest in "market"-based solutions to the climate problem (Chapters 1 and 5), community is becoming an important site in the governing of climate change. A variety of ways in which state and non-state actors seek to engage the community, from education campaigns, to participatory processes, projects, and social movements, can now be found. We find that such initiatives have been driven by attempts to implement policy, by a rationale that seeks to mobilize "responsible" communities into governing climate change outside of the direct control of the state, and also by actors seeking to create an alternative politics of climate change. Community-based initiatives, unlike other forms of climate governance, have succeeded in generating a degree of public engagement with the issue of climate change. However, they are not without challenges in terms of effectiveness, legitimacy, and equity, and it is to these issues that we now turn.

The first issue in terms of the impact and effectiveness of community-based climate governance relates to the different levels of success in engaging the public which can be found across different approaches. Attempts to shift behavior through education campaigns have been found to be largely ineffective, while, on the other hand, community-based projects—such as those discussed here in terms of forestry and renewable energy—have had a much higher degree of success in terms of the number of participants and projects created. This suggests that those schemes which seek to target individual members of the community may have less impact than those that focus on the community scale and seek to achieve change at this level of social organization. It may also be the case, as with the example of community forestry, that the potential material benefits that accrue to communities willing to participate in a project encourage them to do so. In terms of the impacts of the initiatives themselves, each has been successful in relation to some goals, for example in the spread of information about climate change and peak oil (TT), in terms of the promotion and development of demonstration projects (UK renewable energy), or in terms of generating income (Mexico's small-scale forestry projects). However, the impacts of these initiatives and others like them, in terms of reducing GHG emissions and achieving other sustainable development benefits, are less clear. First, such schemes remain—perhaps necessarily—small in scale and their effects highly uneven. Second, accounting for emissions savings and their additionality is challenging, given the range of factors that may be shaping individual behavior, investment decisions, and the emergence of new forms of energy technology. Third, there is evidence that some of the progressive goals of such initiatives (of community involvement, in the case of renewable energy in the UK, or of wider sustainable development objectives in the case of small-scale forestry) are being neglected in state-led community governance initiatives, suggesting that such schemes remain predominantly a means through which the state seeks to further its own interests in terms of climate governance, rather than providing arenas in which community goals can also be pursued. Community-led initiatives, such as the TT example used here, may be more able to retain the broader goals of social and environmental justice that frequently provide the basis for their development in the first place. However, this raises questions as to whether their impact will remain confined to the margins of politics, with little impact on mainstream approaches to climate governance or the vast bulk of GHG emissions. By their very nature they are voluntary, and so communities—if they chose—can opt to do nothing in the absence of pressure to do otherwise.

In terms of accountability, community-based initiatives appear, at least on the surface, to offer a genuine means through which the public can be engaged on the issue of climate change. However, challenges arise in terms of how responsibility for addressing climate change is orchestrated in such schemes, and the extent to which there is potential for genuine engagement in the framing and implementation of climate policy. Rather than being a means through which publics are able to hold those in authority to account for climate change governance, community-based initiatives could be seen as a means through which state (and non-state) actors seek to shift responsibility, and with it accountability, to individuals, households and particular places. This may serve to shift accountability from actors which preside over significant sources of GHG emissions and with the capacity to reduce vulnerability— such as nation-states and large corporations—to those who have little in the way of power to address either the causes or consequences of climate change—for example forest-based peoples in Mexico or rural communities in the UK. Moreover, even where community-based schemes enable external authorities to be held to account for their actions on climate change, the limited nature of participation in such schemes, coupled with restrictions on what is "open" for negotiation, can reduce the accountability of climate change governance. For example, in Mexico changes in rules on carbon sequestration accounting, or in the UK on the funding available for renewable energy projects, were beyond the remit of the communities involved, meaning that the answerability of central authorities to the communities involved in these schemes was limited. In this manner, the use of "community-based" projects by external authorities could function as a smokescreen for other agendas (e.g. the promotion of wind energy in the UK despite "community" protests). Initiatives which have emerged at the grass-roots level potentially have accountability built in, through processes of direct participation which allow people to engage or not, but here the key accountability challenge is one of enforceability—reliant on voluntary participation, they may fall short of achieving their ambitions.

These issues of accountability are intimately connected to those of equity in community-based initiatives. At issue here is how, on what grounds and by whom, the terms of community engagement are established, and with what consequent implications for the distribution of costs and benefits. In particular, the case studies reviewed in this chapter suggest that there are profound equity implications that arise from the interplay of international and national policy agendas with local projects for addressing climate change. For example, in the case of Mexican CDM projects, shifts in the requirements for carbon-accounting externally had significant implications for the social and environmental benefits of forest carbon sequestration. At the same time, seeking to address climate change in certain places may have inequitable geographical impacts, as, for example, benefits from carbon forestry accrue to some communities and not others, some places experience the advantages (or disadvantages) of renewable energy generation, while the localization of food production, promoted by TT, has the potential to create negative implications for farmers in the global South. Community-based initiatives to address climate change are, therefore, not free from the politics of interest, and their development creates both winners and losers, as with all the aspects of climate governance that we consider in this book.

In this chapter, we have suggested that community is becoming a key site in the governing of climate change—a process that warrants attention in its own right, but that also has potentially significant implications for how we conceive of (global) climate governance. In regarding community as a site of climate governance, we have suggested that there is a need to reconsider the way in which the state operates. Rather than being confined to particular institutional boundaries, we have suggested that the state seeks to extend its reach through the transfer of responsibility for climate governance to sites of community. This in turn suggests that the boundaries between the state and non-state arenas in climate governance are, at the very least, blurred, raising questions about the usefulness of such categories in the analysis of climate change governance. However, this extension of the state, and the multiple means through which climate change is being framed as a governance problem within community-based initiatives, raises questions about the limits of climate governance. While, as we outlined in the Introduction, a conception of global climate governance which confines itself to the global scale and to state-based institutions is no longer adequate, the question could also be raised as to whether everything—from a community garden project, to an isolated wind turbine—should be considered as evidence of climate governance. In this context, determining why, and with what implications, particular initiatives are framed as a response to climate change, becomes ever more important. What is clear, however, is that alongside a growing "carbon economy" (Chapter 5) we are witnessing the emergence of a "carbon society" in which projects and initiatives of all shapes and sizes are increasingly justified in the name of climate change.

5 The private governance of climate change

In this chapter, we focus on the increasingly important role of the private sector in the governance of climate change. As market and political actors in their own right, private sector organizations, and key industries in the energy and transport sectors in particular, have played a significant role in the evolution of the climate regime: providing expertise, lobbying governments and international institutions, and traditionally cautioning against strident action on the issue. More recently, however, many firms have come to see action on climate change as a potential opportunity rather than a threat. From emissions trading to the carbon offset market, from investments in technology to the use of a range of corporate social responsibility tools and investment strategies, financial and business actors now play a central role in the day-to-day governance of climate change. We explore how, as well as influencing traditional forms of state-led international governance, private actors have increasingly been constructing their own forms of governance, and the challenges this creates.

Business and climate change

It is difficult to overstate the importance of the private sector in proposals to address climate change. As the Business Council for Sustainable Development acknowledged at the time the UNFCCC was negotiated in 1992, "Industry accounts for more than one third of energy consumed world-wide and uses more energy than any other end-user in industrialized and newly industrializing economies."[1] To reinforce the point, Greenpeace International showed some years later, in a comparison of CO_2 emissions from the burning of fossil fuels by major oil companies with country emissions from fossil fuel combustion, that Shell emits more than Saudi Arabia, Amoco more than Canada, Mobil more than Australia and BP, Exxon and Texaco more than

France, Spain, and the Netherlands.[2] More recently, in 2007, the UN estimated that 86 percent of global investment and financial flows required to tackle climate change will need to come from the private sector.[3] It is not just their contribution to the problem and role in funding solutions, however, which makes business such as key actor. As the International Chamber of Commerce put it:

> Industry's involvement is a critical factor in the policy delibera-tions relating to climate change. It is industry that will meet the growing demands of consumers for goods and services. It is industry that develops and disseminates most of the world's tech-nology. It is industry and the private financial community that marshal most of the financial resources that fund the world's eco-nomic growth. It is industry that develops, finances, and manages most of the investments that enhance and protect the environment. It is industry, therefore, that will be called upon to implement and finance a substantial part of governments' climate change policies.[4]

Business has hence been cast simultaneously as the problem and the source of solutions to climate change. Simply put, it is perhaps truer now than ever that constructive business engagement with the climate issue is a precondition to effective action.

Shaping the climate regime

For a long time, businesses mobilized themselves to stall action on cli-mate change. Climate change first registered on the radar screen of fossil fuel firms in the late 1980s and early 1990s amid growing demands for an international treaty to address the issue. The IPCC had been set up in 1988 and negotiations were set in train through the Intergovernmental Negotiating Committee for a UNFCCC (as we saw in Chapter 1). It was at this moment that businesses, whose interests were threatened by the prospect of action to reduce the use of fossil fuels, began to organize themselves politically. They moved into gear to hire lobbyists and form coalitions to present and advance their inter-ests. Two in particular stood out at this time; the Global Climate Coalition (GCC) and the Climate Council. Members included all the main oil companies, car manufacturers, steel firms, and US electricity generators, as well as Associations like the US National Association of Manufacturers and the National Mining Association.[5] Though now disbanded, it is difficult to overestimate the importance of the GCC in the first half of the 1990s as the dominant voice of concerned industry in the international climate negotiations. As Robert Reinstein, former head of the US climate delegation, and industry lobbyist at the time, told us back in 1996: "when GCC, which represents companies constituting a very significant proportion of the country's GDP start making noises, they obviously get attention."[6] At the national level, groups such as the Confederation of British Industry (UK), the World Coal Institute, the American Petroleum Institute or Western Fuels Association (USA), as well as regional groupings such as employers' organizations (Union of Industrial Employers' Confederations in Europe—UNICE) (now Business Europe) and business round tables such as the European Round Table of Industrialists (ERT), sought to fight off policy measures that threatened their interests. One such measure was the European Community (EC, forerunner of the EU) carbon tax proposed in 1992 which, according to *The Economist*, "spurred the massed ranks of Europe's industrialists to mount what is probably their most powerful offensive against a European Council proposal."[7] Shocked at the intensity of business mobilization against the tax, Carlos Ripa de Meana, EC environment commissioner at the time, described the lobbying offensive as a "violent assault."[8]

A multi-pronged political strategy was developed to promote the voice of industry in climate debates cautioning against hasty action. First, groups sought to challenge the science behind climate change. Scientists such as Fred Singer, Robert Balling, and Patrick Michaels, skeptics of the prevailing consensus on climate change, were provided with funds from fossil fuel companies to support their research which raised questions about human contributions to climate change. Such a strategy was successful in part because the fossil fuel lobby was (and in many ways still is) so well linked to those governments that contribute most to global warming. In the USA, Chief of Staff John Sununu, a well-known climate change skeptic, played a key part in ensuring that President Bush Snr. was exposed only to doubts about the science and studies that emphasized the costs of taking action. Sununu's director of communications in the White House was John Schlaes, who went on to head the Global Climate Coalition leading the business challenge to a treaty on climate change. Insiders such as Sununu certainly made the job of industry lobbyists a whole lot easier.

As the links between climate skeptics and their industry backers became more public, a second strategy started to appeal to those businesses opposing action. Recognizing that skeptical publics were more likely to trust NGOs over government and business in environmental debates, the idea was to create business-funded environmental NGOs or what many critics call "astro-turf" organizations.[9] Their role was to persuade an increasingly anxious public that fears about climate change were misplaced, and to emphasize the crucial role of fossil fuels in economic development. In one case, the Western Fuels Association, the Edison Electric Institute, and the National Coal Association of the United States created the benign-sounding Information Council for the Environment, which launched a $500,000 advertising campaign "to reposition global warming as theory, not fact."[10] This sum was more than the combined amount spent by the Environmental Defense Fund, the Sierra Club, Natural Resources Defense Fund, World Wide Fund for Nature (WWF), and the Union of Concerned Scientists on their climate campaigns. Other examples included the Coalition for Vehicle Choice, a corporate front organization which ran an infamous campaign with a woman complaining that "the government wants to take away my SUV."[11] In the run-up to Kyoto, these groups spent $13 million in an advertising campaign designed to discredit both the Kyoto Protocol in particular, and fears about climate change more generally.

The third element of the strategy was to emphasize the economic costs of tackling climate change. The fossil fuel lobby set to work producing economic studies suggesting that economies would be driven into recession if they adopted measures proposed by leading scientists. The Australian Bureau of Agriculture and Resource Economics (ABARE), for example, conducted a series of modeling exercises that predicted the loss of jobs and income if emissions reductions targets were accepted. The membership of the group's steering committee included Mobil, Exxon (now Exxon-Mobil), Texaco, and Statoil. ABARE's publicity material made it clear why participation at $50,000 was a good deal: "by becoming a member ... you will have influence on the direction of the model development."[12] While the modeling activities of ABARE were submitted for investigation by the Australian Conservation Foundation to the Commonwealth Ombudsman and found to be open to accusations of exclusivity and bias, the key message that acting on climate change would have significant economic costs had a direct bearing on the position adopted by the Australian government during the international negotiations.[13]

Related to concerns about the economics of addressing climate change was the argument that if emissions were only reduced in the North, businesses would merely relocate the most polluting parts of their business overseas, producing a loss of competitiveness and no net gains in emissions reductions. In a globalized economy of capital mobility and internationalized structures of production, it was argued, there is no point in one country taking action if others do not do the same. This has become known as the problem of "carbon leakage."

This led to the adoption of a fourth strategy: double-edged diplomacy aimed at creating stalemate in the negotiations towards an agreement at Kyoto. For example, while the former chief executive of Exxon, Lee Raymond, was busy arguing in the United States that no action should be taken there unless China, India and other developing countries also undertook actions, in front of audiences in China party leaders and their industries were encouraged to resist calls from the United States to take action on climate change since this was a problem that China had contributed very little to. Using domestic politics to stall international progress on a deal was the fifth tactic in the armory of those firms intent on stalling the development of the climate regime. Press conferences at the climate negotiations were used to parade petitions from US senators promising to veto any Kyoto deal that did not include emissions reductions targets for developing countries. This demand came from the Byrd-Hagel resolution passed in 1997 (see Chapter 2) which made US support for Kyoto conditional on developing countries accepting binding emissions reductions obligations, which was a distant prospect at the time. This effectively tied the hands of Clinton's negotiators, since the Senate has to ratify by two-thirds any treaty that the government signs up to.

Each of these strategies fed into the sixth pillar of firms' political strategy at the time: directly influencing the climate change negotiations. Groups such as the Global Climate Coalition and the Climate Council, and in particular their respective heads John Schlaes and Don Pearlman, worked closely with oil-exporting states whose interests were also threatened by the prospect of emissions cuts. From drafting paragraphs of text that delegations would then introduce as their own into draft UN treaty text (often even left in the lobbyists' own handwriting) to hosting press conferences warning of the calamitous consequences of taking action on climate change, the fossil fuel industries managed to gain an important hold over the emerging climate regime. For example, a proposal submitted by Saudi Arabia in 1994 that protocols to the convention should be adopted by three-quarters of the parties instead of the existing two-thirds, was widely attributed to John Schlaes and Don Pearlman, while the World Coal Institute, alongside OPEC and Australia, was instrumental in introducing language ensuring that implementation of the UNFCCC would be done in such a way that account is taken of the impacts on developing countries whose economies are "highly dependent on income generated from the production, processing and export and/or consumption of fossil fuels and associated energy-intensive products."[14] When their interventions and presence on the floor of the UN negotiating halls reached what was considered an un-acceptable level for groups that were officially classified as "observers" of the process, a ban on non-governmental representatives from being present on the floor of the main plenary was introduced, given that these were after all meant to be primarily intergovernmental negotiations.

Private governance

From a long history of opposition to action on climate change, described above, climate change has been repositioned by some elements within the business sector as a business opportunity, which has led to more positive engagement with climate governance initiatives. In this new era, and beyond attempts to influence the international climate regime, we have witnessed the emergence of what is sometimes referred to as private governance in the climate change domain.[15] Private governance can take several forms, and operate either transnationally—as we saw in Chapter 3—or within national or local territories. One type of private governance involves internal processes of self-regulation—the voluntary actions of firms in reducing their own emissions. A second involves the creation of new sites of climate governance through the creation and operation of carbon markets as one way in which actors can fund emissions reductions above and beyond the climate regime. A third involves new forms of private regulation through codes of conduct, standards, and forms of certification, usually in conjunction with other non-state actors, to regulate emissions of GHG from business activities and to address concerns about the credibility of voluntary carbon markets. This includes efforts to improve performance evaluation and forms of civil regulation whose aim is to hold business to account.

Self-regulation

The emergence of a positive approach to addressing climate change among some in the business sector in part reflects growing public concern with the issue and the rise of climate change as a corporate social responsibility (CSR) issue as well as a growing recognition of the economic possibilities heralded by low carbon technologies and new carbon markets. A number of NGOs, such as the Climate Group in the UK and the Pew Centre in the United States, have played an important role in this respect in making the business case for action on climate change and publicizing the benefits achieved by existing leaders in the field.

The extent to which businesses have adopted a positive approach to climate change depends on the region and firm in question.[16] There is, nevertheless, already a great deal of evidence of companies taking voluntary action on climate change to reduce their own emissions, capitalizing on the economic savings to be made and the public relations credit to be earned from being seen to take a lead on the issue. Examples of this type of voluntary governance or self-regulation include chemicals giant Du Pont, which reduced its emissions by 65 percent below their 1990 levels, while IBM saved $115 million since 1998 through cutting its carbon emissions. Meanwhile in 2005 General Electric announced 'Ecomagination,' an initiative that would see the company double investment in research for cleaner technologies to $1.5 billion a year by 2010, double sales of environmentally friendly products to at least $20 billion by 2010, and reduce GHG emissions by 1 percent by 2012 (from 2004 levels). The company is being rewarded financially, with revenues from the company's portfolio of energy efficient and environmentally advantageous products and services exceeding $12 billion in 2006, up 20 percent from 2005, while the order backlog rose to $50 billion.[17] Some firms such as BP and Shell have gone so far as to establish their own intra-firm emissions trading systems which encourage competitive reductions between different parts of the firm. This carries benefits such as saving money through reduced use of energy and first-mover advantages that come from developing new technologies and production processes to meet the targets. They also represent interesting governance initiatives in their own right, emerging as they did ahead of even the European Union's own emissions trading scheme.

In this context of a growing interest in self-regulation, many firms have even come out in favor of a binding regulatory framework to lay out clear emissions targets for the immediate and long-term future—in part to protect the competitive advantages that so-called first movers on GHG emissions reductions might achieve. In 2007, an impressive group of 150 of some of the world's best known companies (including Volkswagen, Shell, Nokia, Kodak, Philips, HSBC, General Electric, Nestlé, Adidas, Nike, Rolls-Royce, DuPont, and Johnson & Johnson) published a "Communiqué on Climate Change" in the *Financial Times* calling for "a comprehensive, legally binding United Nations agreement to tackle Climate Change" which included the following statement:

> The shift to a low carbon economy will create significant business opportunities. New markets for low carbon technologies and products worth billions of dollars will be created if the world acts on the scale required. In summary, we believe that tackling climate change is the pro-growth strategy. Ignoring it will ultimately undermine economic growth.[18]

Constructing carbon markets

Above and beyond positive support for the climate regime, which marks a significant shift from the oppositional role that the majority of firms once occupied in climate governance, some companies have sought to create their own carbon markets outside of the international regime and which go beyond the CDM we discussed in Chapters 1 and 2. Here we will mention two areas where firms have created their own forms of voluntary climate governance through the market: the carbon offset market and climate exchanges.

First, carbon offset companies were set up to meet increasing demand from companies and individuals to fund projects and activities aimed at reducing GHG emissions, which could be used to offset everyday activities or claimed against voluntary targets in the case of firms. Many firms set themselves up as intermediaries between project developers and the "sellers" of emissions credits, often based in the global South and project "buyers" in the North. Making this market exchange possible are the verification firms (like Société Générale de Surveillance (SGS), Det Norske Veritas (DNV), and Tüv Süd) who assess the project in terms of criteria set out by agreed standards which allows credits to be issued to the buyer. Some firms operate as wholesalers, such as Eco-Securities, buying and selling on credits to other firms such as Climate Care. Climate Care focuses more on the development and retail of voluntary and compliance carbon offsets, handling anything from individuals' offset requests to serving the needs of corporate clients like British Airways and Land Rover. Eco-Securities on the other hand has become one of the largest firms working in the compliance market. It has registered more than 110 CDM projects and has more than 400 underway in 34 countries, with a portfolio of projects that will total more than 122 million CERs by 2012. With offices in the UK, USA, Indonesia, and India, it is one of the key global players. When it was floated on the London Stock Exchange it raised €80 million and in 2007 raised a further €44 million with Credit Suisse acquiring almost 10 percent of the shares.[19]

The importance of voluntary carbon markets is increasing, with a tripling of transactions between 2006 and 2007, for example. Project-based carbon offsets could be worth at least €200 billion by 2020. Though other actors are involved, businesses are buying nearly 80 percent of credits (50 percent to offset emissions, 29 percent for investment/resale). NGOs account for just 13 percent of demand, individuals account for 5 percent of the market, while governments are only responsible for 0.4 percent of the purchases. Reflecting the business nature of the market,

25 percent of total traded volume is used to directly offset emissions; 75 percent changed hands and could be resold in the future. That said, price and convenience are cited as the least important factors driving actors' engagement with voluntary markets. Considerations of additionality, certification, reputation, and environmental and social benefits are cited as most important. This reflects the fact that CSR and public relations are offered as the most common motivation behind offset purchases. To put the importance of voluntary carbon markets into overall perspective though, they remain only fraction of the size of regulated carbon markets at 2.2 percent.[20]

A second means of creating climate governance through markets was the construction of a private site of carbon exchange.[21] Work on the Chicago Climate Exchange (CCX) started in 2000 and it was launched in 2003. It is a member organization, where large corporations who join agree to reduce their greenhouse gas emissions as a condition of membership. All members who have joined to date have agreed to reduce their emissions by 6 percent by 2010 (roughly tracking the Kyoto target, although the baseline year is not 1990). The Chicago Climate Exchange describes its commitments as "voluntary legally binding commitments."[22] In the CCX member firms can meet their target through internal reductions, by purchasing credits on the exchange from other members or through offset projects organized through the exchange. The reduction in carbon emissions must be made in North America, however, and verified by an independent third party. The decision-making process is open to all members that abide by the rules of the exchange, and six committees set specific technical rules and standards.[23] The CCX is based in Chicago, and started with just US and Canadian firms (like Ford, DuPont, and Motorola), but now has firms participating from other countries, notably Australia, China, and India, as well as some government agencies from Brazil and the Indian embassy in the United States. It has also established the European Climate Exchange, which operates as a registry and exchange in the EU ETS, and more recently, the Montreal Climate Exchange, which started futures trading in GHG emissions allowances in June 2008. Important in terms of the globality and effectiveness of this form of climate governance, it also now has an affiliated carbon market in China in the form of the Tianjin Climate Exchange.

In addition to creating "private" markets for carbon governance, businesses have also sought to shape the broader agenda surrounding the nature of carbon markets. Just as fossil fuel interests organized themselves into coalitions to influence the climate regime, business associations have also been created to attempt to shape the rules which govern the carbon economy. Bodies such as the International Emissions Trading Association (IETA) have developed a self-interest in protecting and expanding a thriving carbon market in which they and the firms they represent can make money. They have been active at the national and international level, engaging in debates about the criteria for inclusion and assessment of potential projects, and promoting voluntary standards which seek to maintain the credibility of voluntary markets in the face of criticism from NGOs and others about the authenticity of the claimed emissions savings amid accusations of "climate fraud."[24] In so far as forms of private governance are unable to deliver their promise or go beyond compliance, however, there may be pressures to (re)regulate to ensure universality, consistency and equity, and to make use of the sanctioning powers which the state alone has access to. For example, further scandals about climate fraud in voluntary markets may lead to demands for government-led guidelines such as the Code of Best Practice produced by the Department for Environment and Rural Affairs (DEFRA) in the UK or even tougher binding standards. In early 2008 DEFRA made a controversial decision to only accredit compliance offsets (CERs) for use on the UK voluntary offset market, a move which encouraged organizations such as Climate Care to source and invest in compliance market projects.

With such pressures set to intensify, there may, over time, be a race to quality as higher demands are placed on voluntary projects and the standards they abide by. As much as 50 percent of the transactions conducted in 2007 involved credits verified to a specific third party standard. According to the report, *State of the Voluntary Carbon Markets 2008*:

> Over the past two years numerous writers and analysts have likened the voluntary carbon markets to the "wild west." In 2007 market trends highlight that this frontier has become a settlement zone. Customers are increasingly savvy about the opportunities and pitfalls in the carbon offset domain and stakeholders are aggressively working to forge the rule of the game and structures to enable smooth transactions.[25]

Rather than viewing them in conflict, the director of Climate Care, Mike Mason, talks about voluntary markets and initiatives "topping up" the compliance markets. Perhaps that is the best that can be said of their role:[26] filling stop-gaps, plugging holes, and engaging and enrolling actors that would otherwise not be interested in taking action. Interestingly though, this might include bringing on board governments that chose to stay out of Kyoto. Firms that see the potential in carbon markets in countries that until recently were outside the Kyoto regime, such as the United States and Australia, have provided a big market for offset providers. Offset firms speak confidently of the ability to "engineer a shift in political stance" of some key players that had been opposed to action, once they see the money to be made.[27]

Private regulation

In response to concerns about the credibility of projects and credits traded in voluntary markets, and the extent to which they genuinely deliver social and environmental benefits, businesses have sought to set up their own forms of standards and voluntary regulation that allow those firms that want to, to show that they are compliant with certain performance criteria. Examples of private governance as standard setting and certification include the CDM Gold Standard, the Voluntary Carbon Standard, and the Climate, Community and Biodiversity standards. They are significant in governance terms because of the steering roles they perform and the informal forms of regulation and standard setting they generate. A number of these standards claim to have at least as stringent criteria for measuring additionality as the CDM. The Gold Standard, initiated by WWF International, Helio International, and South South North in 2003, includes among its objectives helping to boost investment in sustainable energy projects and increasing public support for renewable energy and energy efficiency.[28] The Gold Standard essentially applies an extra set of screens to CDM or voluntary projects using strict additionality criteria and certifying with Gold Standard credits only those projects in the areas of renewable and energy efficiency and methane-to-energy. To encourage sustainable development, it also places emphasis on local stakeholder consultation prior to implementation. The "boutique credits"[29] that result from these extra transaction costs are generally sold at about 25 percent above the market value for normal CERs. In many ways then it is a "beyond compliance" initiative, as it is designed to help the CDM realize its objectives and is fully integrated with CDM procedures; so, for example, verifiers (Designated Operational Entities) will validate both standard CDM and Gold Standard requirements at the same time.

The VCS, developed by the Climate Group, the IETA, and the World Business Council on Sustainable Development (WBCSD) in 2006, seeks to provide a "robust global standard, program framework and institutional structure for validation and verification of voluntary GHG emission reductions."[30] It aims to "experiment and stimulate innovation in GHG mitigation technologies, verification and registration processes that can be built into other programs and regulations."[31] Part of this involves performing key governance functions such as guarding against double-counting of the same emissions reductions and providing transparency for the public. The VCS board comprises nine members from across the public and private sectors, including the World Economic Forum, IETA and International Institute for Environment and Development (IIED), while its steering committee is made up of many of the key actors in this area discussed above, including DNV, Eco-Securities and Cantor CO_2e. By providing rules for certification that are claimed to be as robust as those of the Kyoto Protocol's CDM, the VCS aims to give confidence to buyers and sellers. As the Climate Group put it, it does this by providing a "set of criteria that will provide integrity to the voluntary carbon market."[32] Two recent reports have found that the VCS is already the most popular standard for voluntary offset projects,[33] is leading the way by being the first carbon standard to cover all major land use activities (whether forestry or agriculture) and that it addresses many of the "permanence and additionality" concerns that have held back the take-up of projects in global carbon markets.[34]

Besides issues of environmental integrity, some standards address the social dimensions of voluntary projects. The CCB standards,[35] for example, released in 2005, aim to "identify land-based carbon projects that deliver robust GHG reductions while also delivering net positive benefits to local communities and biodiversity."[36] These provide gold, silver and standard certification for projects, depending on how many of their criteria are satisfied. These range from required criteria which include additionality, baseline issues and leakage assessment, through to extras such as employing stakeholders in project management, worker safety and adaptive management. Given the fatigue caused by the proliferation in voluntary standards, one thing CCB does is build on the use of existing certifiers, authorized under Kyoto or by the Forestry Stewardship Council for example, as third party evaluators of whether a project deserves to be certified.

Other initiatives that fall under the heading of private governance and which use forms of certification and benchmarking are about promoting the transparency and accountability of investors. By recording, evaluating, and comparing performance they may also, however, create pressures for firms to reduce their emissions through their investments. The Carbon Disclosure Project (CDP), for example, by systematizing information about investor's emissions, creates the means to pressure firms to invest in renewable rather than fossil fuel energy solutions. The

CDP now covers US$57 trillion worth of assets from over 3,000 companies. The scope of private regulation is, therefore, impressive and reaches key actors not subject to other forms of governance. It claims:

> The CDP provides a secretariat for the world's largest institutional investor collaboration on the business implications of climate change. CDP represents an efficient process whereby many institutional investors collectively sign a single global request for disclosure of information on Greenhouse Gas Emissions. More than 1,000 large corporations report on their emissions through this web site. On 1 February 2007 this request was sent to over 2,400 companies. 1,550 companies and 385 institutional investors from different regions and sectors participated in and complied with the CDP's 2007 questionnaire.[37]

The Greenhouse Gas Protocol (GHG Protocol) is a corporate reporting and accounting standard, jointly created by the World Resources Institute (WRI) and the WBCSD in 1998 and now claims to be "the most widely used international accounting tool for government and business leaders to understand, quantify, and manage greenhouse gas emissions."[38] Emphasizing the links between formal and informal regulation, in 2006 the International Organization for Standardization (ISO) adopted the Corporate Standard as the basis for its ISO 14064-I: Specification with Guidance at the Organization Level for Quantification and Reporting of Greenhouse Gas Emissions and Removals. On 3 December 2007, ISO, WBCSD and WRI signed a Memorandum of Understanding to jointly promote both global standards.[39] Organizations such as the Carbon Fund meanwhile aim at: "increasing awareness of products and companies that are compensating for their carbon footprint while helping to hasten a market transformation."[40] Indeed, tools such as the Carbon Fund's "Carbon Footprint Protocol" draw on guidelines and standards that govern the compliance market, such as rules for CDM and Land Use, Land-Use Change and Forestry (LULUCF), since essentially they are wrestling with the same issues of proving additionality and using valid baselines.

As with all forms of climate governance, private climate governance has been contested by other actors—disputing the claims companies have made about their performance, raising concerns about the adequacy of voluntary regulation of markets, and seeking to develop tools and strategies of accountability. Such strategies are sometimes referred to as civil regulation: civil society-based regulation of the private sector.[41] Civil regulation includes the use of shareholder activism, consumer boycotts, and exposé campaigns.[42] Alternative systems of sanctions to penalize irresponsible conduct and reward positive behavior include loss of market value or consumer confidence, tarnished public reputation and disaffection among shareholders that serve to "harden the environmental accountability demands leveled at corporations."[43] Not only do they provide "an instrument of accountability for ecological performance," a critical dimension of their effectiveness derives from the construction of these mechanisms of civil redress.[44] Corporations are, through these means, subject to a variety of accountability sanctions that go beyond the strict public regulation of their activities.

To take one example, activists have increasingly targeted the power and vulnerability of banks, trying to persuade them that investments in fossil fuels should increasingly been seen as liabilities rather than assets. Many leading firms have been subject to shareholder activism. The year 2005 saw a record number of shareholder resolutions on global warming. State and city pension funds, labor foundations, and religious and other institutional shareholders filed 30 global warming resolutions requesting financial risk and disclosure plans to reduce GHG emissions. This is three times the number for 2000/01. Firms affected include leading players from the automobile sector such as Ford and General Motors; Chevron Texaco, Unocal, and Exxon Mobil from the oil sector; Dow Chemicals, and market leaders in financial services such as J. P. Morgan Chase. Groups such as CERES (Coalition for Environmentally Responsible Economies) and ICCR (Interfaith Centre for Corporate Responsibility), a coalition of 275 faith-based institutional investors, have been using their financial muscle to hold firms to account for their performance on climate change. They demand both information disclosure and management practices that reflect the values of their shareholders. Overall, the resolutions led to distinctive agreements, but with some common links: acknowledging climate change impacts in securities filings and on corporate websites; assigning board-level responsibility for overseeing climate change mitigation strategy; and benchmarking and GHG emissions reductions goals.[45] That many firms are now engaging with the CDP as a means of making their climate-related activities transparent may reflect the increasing pressure that such forms of civil regulation are bringing to bear.

Governance issues and challenges

We can see from the above examples that the role of business in climate politics produces a wide-ranging tapestry of governance practices, taking uneven forms across scales and regions. This is perhaps inevitable, given the range of actors whose behavior we are seeking to change through climate governance, but it does draw our attention to a number of critical features of the climate governance landscape.

One key feature that emerges from this chapter is the continued importance of the relation between forms of "public" and "private" governance. As we saw in the history of business involvement in the climate change negotiations, private actors have had a profound influence on the nature of public climate governance. But the reverse is also true. As is clear from the development of both market and voluntary approaches to governing climate change, the state continues to play a significant role in shaping private climate governance. Abyd Karmali, managing director and global head of carbon markets at Merrill Lynch, puts it succinctly:

> Those who assume that the carbon market is purely a private market miss the point that the entire market is a creation of government policy. Moreover, it is important to realize that, to flourish, carbon markets need a strong regulator and approach to governance. This means, for example, that the emission reduction targets must be ratcheted down over time, rules about eligibility of carbon credits must be clear etc. Also, carbon markets need to work in concert with other policies and measures since not even the most ardent market proponents are under any illusion that markets alone will solve the problem.[46]

We are witnessing the continuation of a dynamic whereby innovative forms of private governance either respond to the limits of public regulation or strive to meet a market niche in doing so, such as the Gold Standard, or seek to anticipate and pre-empt further public regulation which may be less sensitive to their interests, such as in the case of the VCS. In short, the shadow of regulation looms large for some private sector actors as an incentive to get their own house in order before governments threaten to do it for them.

Despite this "shadow of hierarchy," a second feature of the emergence of private forms of governance has been the breadth and scope of initiatives that are underway, often going beyond standards required in international and national policy. A positive reading of this new landscape of private regulation is that it goes further and faster than state-based action on climate change, and creates new channels of pressures on key contributors to climate change. With the rise of CSR and voluntary regulation, social expectations about the responsibilities of firms sometimes far outstrip those that are expressed in legal instruments, where there remains an imbalance between the rights and responsibilities of firms as they are constituted at the global level.[47] Voluntary standards are also quicker to approve, respond to private and consumer needs and potentially provide wide coverage if adopted by large firms with extensive supply chain networks. When a company like the supermarket giant Tesco in the UK announces its intention to ultimately label the carbon footprint of all the products it sells in its supermarkets, the global reach of the measure is significant. A further contribution private governance can make is to improve and systematize levels of reporting and disclosure about GHG emissions, which in itself is a form of governance and may support other standards or forms of regulation. Beyond serving as a source of credible information on GHG offsets, for example, the Offset Quality Initiative founded in 2007 by six non-profit actors (including the Climate Trust, Pew Centre on Global Climate Change, and the Climate Group) seeks to advance the integration of key principles on carbon offset quality in emerging climate change policies at the state, regional, and federal levels.

There is a key difference, nevertheless, between state regulation of and for the private sector and regulations established by and for private actors themselves. Voluntary regulation is discretionary and derives from incentives some leading firms have to demonstrate leadership or to seek to exploit a "first mover" advantage in the market. The problem with relying on such responses is that only those firms under pressure (from within and outside the firm) and in the public spotlight will act. Given the choice, many will not, leading to problems of uneven obligations, and free-riding by those that benefit from the actions taken by others without making sacrifices themselves. This is true within states and internationally, where, in the absence of global coverage, action may be ineffective if key sectors and polluters are not on board, and unappealing for firms if private sectors elsewhere are not also making sacrifices, especially given the problems of "carbon leakage" described above. If we rely on making the "business case" for action on climate change, sectors that make their profits from burning fossil fuels may just not do anything.

A related concern is the issue of effectiveness. If the different mechanisms of private governance discussed here are to make a meaningful contribution to efforts to tackle climate change, they have to show that they can bring down emissions. Yet the coverage of many instruments is limited. The CCX, for example, has so far been unable to attract companies from energy-intensive sectors. Only three of the top 50 *Financial Times* global companies are members of the exchange, and all are from less energy-intensive sectors. In general it seems companies participate more frequently in information-based climate change arrangements than in voluntary or legally binding market-based arrangements, and geographically speaking there tends to be a greater representation of European and North American transnational companies in private governance arrangements. With the CDP, for example, the majority of the institutional investors are based either in Europe or North America. Company participation is often uneven or restricted.

There are also issues of process, of who participates and how, and of accountability for actions and inactions. Some forms of private governance allow for more participation, transparency and accountability than others. The CDP questionnaire and reports, for example, are public and can be accessed via the internet. Responses from companies are available without restriction. The CDP operates on an essentially self-select basis, in so far as the initiative is voluntary. Yet the flexibility built into the CDP means that firms are able to choose which of their operations they include in their emissions disclosure. Indeed, most companies signed up to the CDP place a disclaimer that the information they enclose does not include their activities in some developing countries. Moreover, there is no institutional control mechanism in place to monitor and verify company responses, though the levels of public access noted above do mean other actors are in a position, at least in theory, to challenge or investigate for themselves claims made by firms submitting data. Interestingly, in the case of the CCX, the US Financial Industry Regulatory Authority handles verification and monitoring issues because the organization is primarily a commodity trading body, suggesting once again that private initiatives often fall back on state authority.[48]

Thinking about private governance generates interesting theoretical as well as policy challenges, therefore. The everyday forms of governance and regulation that firms generate, often on scales and overseeing resources far greater than those over which governments have direct control, are poorly captured by traditional explanations of governance which focus on the public, the interstate, the global and the formal. The increasing profile of business in these debates reflects not just the turn towards market-based solutions to climate change which affect and imply business involvement (as we saw in Chapter 1), but also broader shifts in power between state and market in a context of globalization. Among different sectors of business, the key role of finance in contemporary forms of neo-liberalism is particularly notable and explains the focus on carbon disclosure, efforts to woo the insurance industry and to work with leading banks such as HSBC.[49] The forms of governance deployed—the emphasis on partnerships with civil society and other actors, voluntarism and networks—reflect neo-liberal modes of governance and their embrace of non-state forms of governing in particular. More specifically, the way firms draw boundaries around what is to be governed, how, by whom and for whom, are very different from the way states and international institutions or cities do, for example. The ways in which they produce governance raise difficult issues of transparency, representation and legitimacy. And yet whether it is oil firms or banks, business actors are increasingly central to any effort to govern climate change. For all these reasons, private governance forces us to think more innovatively about how and where governance happens and who is served by particular arrangements.

6 Conclusions

Once upon a time studying and keeping track of the world of climate governance was relatively simple. By attending the UN climate change negotiations or reading *Earth Negotiations Bulletin*[1] and the NGO newsletter *ECO* you could keep up to date with what was going on. If you take the view that international institutions continue to be the primary and most important site of climate governance, your task continues to be a relatively straightforward one, even if the range of issues on the negotiating table has grown more complex. If, however, you adopt a broader view of climate governance—one that raises more open questions about how and where it happens and who is engaged in it—as we have done in this book, your task is altogether more complex. In this final chapter of the book we summarize some of the key cross-cutting themes that have emerged throughout the book, what they imply for governing climate change, and how we can make sense of them.

The shifting terrain of governance

Claiming that the landscape of climate governance is more plural, diverse and multifaceted is not to say that the governance of climate change by and through international institutions and nation-states does not continue to be a central part of the story. Key decisions about overall targets and the means of delivering them continue to be made at the international level by nation-states operating within global institutions. We saw in Chapter 1 how historical and contemporary conflicts along North–South and other lines continue to get played out in the UN negotiations through debates about responsibility for the problem and the appropriate responses to it.

However, we have also seen examples of a tremendous proliferation in governance initiatives around the world at all scales of politics (from the local to the global) that enroll a vast array of state, non-state, public and private actors in different dimensions of governing. These have taken the form of public-private and hybrid initiatives (such as REEEP and CDM), of private codes and standard-setting (such as the VCS or Gold Standard) or community-based initiatives such as Transition Town projects. As the dimensions of the challenge of tackling climate change become clearer and the debate moves from setting goals to implementation, it has become increasingly obvious that climate governance can no longer be achieved by the actions and initiatives of international institutions or national governments alone. Rather, the growth of new "sites" of climate politics that we have documented in this book is testament to the increasing mainstreaming of climate change—its integration in other policy domains (biodiversity, trade, energy) and its increasing uptake by a range of organizations (be they corporate, donor or civil society) beyond the realm of what was traditionally considered as climate governance.

We have seen, though, that with this shifting terrain come a number of critical challenges. Governing across so many scales and through so many dispersed but overlapping networks presents huge problems of coordination and policy coherence. We saw in Chapter 2 how the number of banks, donors, and NGOs clamoring to define a leading role for themselves in the delivery of carbon finance is creating problems of duplication and turf wars over who funds what. As a premium is placed on the delivery of clean energy we have seen the REEEP initiative set up, the APP created and numerous energy investments overseen by the World Bank and regional development banks in Asia and Latin America. In such a crowded field, there are challenges of ensuring a rational division of labor which allows each actor to do what it does best.

Just as actors beyond the traditional world of climate policy are becoming increasingly involved in its governance, so too though the nature of climate change as an issue means that it spills over into the governance of other sectors, such as energy, trade, industry, agriculture and housing, to name but a few. This creates problems of policy coherence at the national level; ensuring that actions in such areas are compatible with and do not undermine climate policy goals. But these issue linkages are also apparent at the international level in debates about the compatibility of climate policy measures with the international trade regime, for example. Tentative US and EU proposals to use border tax adjustment measures against imported goods that have been produced in an energy or carbon-intensive manner (such as Chinese steel) could well face a legal challenge in a dispute under the remit of the WTO.[2] How this issue, and others, is resolved will be shaped by who wields the most power and whose rules rule. Managing the shifting terrain and multiple sites of climate governance is not just a matter of improved coordination. The power of key actors with interests at stake will be brought to bear in determining what counts as climate governance, and how and where particular issues such as this get resolved.

New actors, old politics?

The shifting terrain of climate governance is in part a reflection of the realization that the resources, capacity, expertise, networks, and power of actors as diverse as states, firms, cities, communities, civil society organizations, and individuals are required to effectively address all aspects of the problem. While nation-states and international institutions remain critical actors, what we have found is that the delivery of the targets and agreed commitments often implies an important role for many other actors, such as businesses and cities, and in reality is dependent on their support and engagement if it is to be effective. As David Levy argues, "if an agreement cannot be crafted that gains the consent of major affected industries, there will likely be no agreement at all."[3] They are after all the "street level bureaucrats" that will be expected to put regulation into practice.[4] Often, however, as well as helping to realize the goals of the regime through their own independent actions, these actors are going beyond the formal climate regime by setting more ambitious targets than those contained in Kyoto. For example, the 2007 Bali World Mayors and Local Governments Climate Protection Agreement states that municipal authorities will seek to reduce GHG emissions by 60 percent worldwide by 2050, with an 80 percent reduction in emissions from industrialized countries.[5] Non-state actors also engage in forms of governance innovation that precede and predate those adopted by government. The oil giants Shell and BP had both developed intra-firm emissions trading schemes from 2000, for example, three years before the EU ETS was established.

The involvement of multiple actors in the governing of climate change has not been a harmonious and consensual process, however. We have seen how actors with conflicting interests in the debate compete to assert their preferred understanding of the nature of the challenges associated with climate change. We saw in Chapter 1 how there has been fierce contestation over whether sufficient scientific consensus exists to justify potentially costly forms of action. Conflicts over which evidence is highlighted in the policy makers' summaries produced by the IPCC reflect what is at stake in the recommendations that flow from how the risks associated with acting or not acting are presented. We saw in Chapter 5 how the business community throughout much of the 1990s sought to present climate change as a threat to economic growth rather than an opportunity to develop new forms of low-carbon business. Battles continue today among and within firms about whether climate change is a threat or an opportunity to their business.

The issue of who governs directly impinges upon the question of who benefits from current governance arrangements. We saw in Chapter 1 that disputes over whether the Global Environment Facility was the appropriate institution to oversee the delivery of aid and technology transfer to developing countries as part of the UNFCCC agreement reflected concerns that the body is too tightly controlled by the World Bank to be responsive to developing country concerns. We have also seen how many smaller developing countries, those most vulnerable to the effects of climate change in many cases, are poorly represented in the international negotiations and lack the legal and scientific capacity to engage fully in the discussions taking place. We used the example of the legal support given by FIELD, the environmental legal NGO, to the Alliance of Small Island States most affected by sea-level rise to bolster their negotiating capacity vis-à-vis more powerful countries and regions. At the same time, the mutual dependence of state and non-state actors in realizing climate governance goals is leading to compromises and new forms of partnership. In Chapter 4 we saw how various forms of community are being mobilized by states and international institutions in order to address climate change. This can bring benefits, both in terms of addressing climate change and for wider goals of social and environmental justice, but can also lead to the exclusion of those actors' interests and goals that do not fit with particular interpretations of the climate governance "project." Attending to the politics of who governs, and on whose behalf, will be critical to ensuring that addressing climate change does not come at a cost to those least able to represent their interests.

The rise of market and voluntary governance

Throughout the book we have also seen that who governs, and importantly how they govern, is in a continual state of flux in the world of climate politics. What is possible and likely from climate governance is a feature of the health of the economy, the nature of relations between states, and which other issues compete with climate change for resources and attention, all of which are subject to change. This broader political and economic context shapes and is shaped by climate governance. We saw in Chapter 2 how the popularity of market-based mechanisms in climate governance can best be understood within a prevailing context of neo-liberalism: an ideology and set of practices which construct a minimal role for the state and view the market as the main provider of welfare and efficient outcomes. The embrace of emissions trading and the creation of the CDM, where earlier they had been opposed, can be understood as part of the deepening hold of neo-liberalism throughout the 1990s as well as the ability of its most powerful proponent, the USA, to insist that such solutions be part of the global deal. Likewise, the call for communities to govern themselves in order to address climate change, as discussed in Chapter 4, and the growth in partnership approaches, explored in Chapter 3, are also modes of governance that fit with neo-liberal ideas about the respective roles of the state, of the private sector, and of individuals.

The cases in this book have also demonstrated a dynamic relationship between formal and informal forms of soft regulation. The CDM Gold Standard is a private non-governmental initiative that provides "compliance plus" incentives for investments in projects that meet certain sustainable development criteria. It does not compete with the CDM rules agreed under the Kyoto Protocol, but rather provides the incentives and means for those who want to go beyond that to do so. We also saw how voluntary governance experiments, such as with forestry projects in voluntary carbon offset markets, can prepare the way for their acceptance into the formal climate regime. On the other hand, as we saw in Chapter 3, some initiatives are seen as a threat to the climate regime. The APP was interpreted as an attempt by the United States to develop an alternative approach to a global legally binding emissions reductions framework by focusing on clean technology cooperation among a select number of leading states, rather than overall emissions reductions targets.

The normalization of market and voluntary approaches to climate governance as a legitimate and appropriate response to the threat of climate change has, therefore, not been without contest. In particular, with the rise of carbon markets as a means of allowing wealthier countries to meet their emissions reductions obligations, we have seen contestation over their consequences, particularly for poorer communities whose resources (land and forests) become subject to an international trade in emissions reductions (Chapter 2). This has resulted in instances of displacement and conflict that critics refer to as "carbon colonialism"[6] as local regimes of resource management are brought into the realm of global climate governance. Likewise, Chapter 4 showed through the example of carbon forestry projects in Mexico how important local institutions are in ensuring that local communities can capture some of the benefits of engagement with the carbon economy. These examples suggest the importance of the link between the procedural and distributional aspects of climate governance: who participates and who decides exercises a powerful influence over who wins and who loses.

Consequences and contestations in climate governance

Amid such diversity among actors and practices of climate governance, it is difficult to form definitive judgments about what is working and for whom. One answer is that it is difficult to know in quantitative terms who is reducing by how much as more actors get involved in setting their own standards employing different baselines and covering different GHG (as we saw in Chapter 5). As responsibility is diffused and governance tools proliferate, maintaining a clear view of what is being achieved overall is very difficult. Symbolically, the fact that governments advance slowly towards a successor to the Kyoto Protocol, as we saw in Chapter 1, or the fact that many of the world's leading firms now endorse the idea of action of climate change, as we saw in Chapter 5, represents success on one level. Key actors are engaged and the momentum is there.

And yet despite the enormous proliferation of initiatives aimed at reporting, benchmarking, and measuring performance, at funding projects and trading credits, that we have seen throughout the book, it would be difficult to argue that the world is showing genuine progress in moving away from a model of development that is fueling climate change. Continued emissions growth tells a different story. Economic growth and emissions trajectories of GHG continue to be closely aligned and governments and corporations alike continue to scour the earth for new sources of fossil fuels, in spite of full knowledge of the human and ecological consequences of burning them into the atmosphere. Rather than prioritize fundamental changes in production and consumption of energy, the world's most industrialized countries are seeking to locate the lowest-cost ways of reducing emissions by identifying emissions reductions projects in the global South or to make more money by trading their way out of trouble through the buying and selling of emissions permits. We have seen how climate justice movements and critics of carbon trading in particular have sought to question the assumed efficiency and effectiveness of such market-based governance instruments, as well as highlight the inequalities and injustices that they create or reinforce. But in a context of such power inequalities within and between societies, the means of holding to account those that benefit most from what has been called the "un-governance" of climate change by those that are or will suffer the worst consequences of inaction are often just not there.

One conclusion that can surely be drawn from the evidence presented in this book, however, is that the landscape of climate governance is ever changing, and interesting and unpredictable alliances between actors, even former adversaries, are increasingly common. It is to be hoped that enough powerful allies in the world of finance and business can be brought on board alongside enough governments with the will and power to lead on action on climate change, pressured, cajoled and shamed into action by an increasingly active public and civil society, to adequately address perhaps the greatest collective action problem the world currently faces.

Theoretical implications

How then are we to explain and make sense of the patterns of governance described above? In the Introduction we provided an overview of some key themes and literatures that we believe are useful in explaining this complex tapestry of climate governance. Here, we gather some of the key insights that emerge from this book, having now examined a range of forms of climate governance.

If we accept that climate governance now takes place in many more places and is produced by a greater range of actors, it is clear that we need to adopt theoretical tools and concepts that allow us to go beyond the state as the primary focus of analysis. Emphasizing the role of non-state actors and explaining their influence on international climate policy is one important aspect of this.[7] But given that many of these actors are cooperating and working together in ways which bypass the climate regime altogether, we need to explore their roles as governance actors in ways that are not defined by whether and when they affect state policy. We saw in Chapter 3 how organizations such as the Climate Group have been able to bring together novel coalitions of cities, businesses, and civil society organizations to share and build on successful initiatives to reduce GHG emissions. An emphasis on networks that cut across divisions of national and international is useful here.[8] We also saw how work on public-private partnerships can also usefully be applied to governance innovation in the climate arena.[9] A key challenge that remains is how to account for the fact that some partnerships are more successful and more powerful than others.

The political economy approaches that we discussed in the Introduction offer promising ways forward for understanding the relations of power that underpin political coalitions which seek to deliver or frustrate change. They emphasize in particular the role of economic actors in the very processes which are subject to regulation, where powerful interests are at stake and from whence change ultimately has to come. By focusing on the dependencies and material and political alliances that bind together state, market, and civil society actors, they provide a sense of how much autonomy, "policy space" and scope for effective action they have.[10] This could be the developing country governments dependent on World Bank finance, or developed country governments reluctant to regulate the emissions of businesses that threaten to relocate their operations elsewhere. But it could also be the coalitions formed between financial capital in the city of London and NGOs pushing business to commit to action, or conflicts between different business interests which create openings for change. This provides us with a useful way to understand why climate is governed as it is and on whose behalf, by placing power at the heart of governance analysis.

All of this implies broader understandings of power than have traditionally been used in the study of climate governance. Notions of "power over" and the ability to demand actions of others continue to be important: the power of states over firms and local authorities is critical to their ability to ensure they enforce actions on climate change. But many of the forms of private governance that we discussed in Chapters 3 and 5 show that even without sanctions, forms of peer and consumer or activist pressure often provide incentives for compliance.[11] We have also seen how power operates in more subtle ways, through the presentation of knowledge; the inclusion or exclusion of emphasis on some aspects of climate science and economics and not others. Appeals to moral authority are strongly made by developing countries, as well as underpinning NGO-led proposals on contraction and convergence or the greenhouse development rights framework that was discussed in Chapter 1. Considering power in these terms means understanding how it works through discourse and in the everyday practice of policies and actions to address climate change.

The power that actors wield in climate governance is not just visible within the self-proclaimed arenas of climate policy, nor does it derive primarily from the power they wield in that area, however. As Susan Strange argues in her critique of regime theories of international governance:

> since the chain of cause and effect so often originates in technology and markets, passing through national policy decisions to emerge as negotiating postures in multilateral discussions, it follows that

attention to the resultant international agreement of some sort is apt to overlook most of the determining factors on which agreement may, in brief, rest.[12]

Likewise, the World Bank is an increasingly central actor in climate governance, not primarily because of its role in developing the Prototype Carbon Fund or the Climate Investment Funds, but because it exerts such influence over the overall development strategies of so many developing countries. Separating its role as the world's largest lender from its role as one among many actors in climate governance makes little sense, because how developing countries engage with it as a climate actor is shaped by the knowledge of the power the institution wields over all other aspects of their economies.

Climate governance is ironically both a microcosm of a larger global political economy, but also a meta-feature of that system in so far as virtually all areas of political activity have an impact on, or might be understood as forms of, climate governance. For instance, the WTO or the World Bank would not consider themselves environmental organizations or climate agencies, but the mandates they have and the influence they wield mean they have a tremendous impact on the level of GHG emissions that pass into the atmosphere. Trade agreements that mean goods are transported over larger distances and financial loans for coal-fired power stations in the case of the World Bank make the work of the official climate regime that much harder. Ultimately, whether we have the collective capacity and will to address climate change in the time frames available to avoid its most dangerous consequences will depend on fundamental change in policy areas beyond the direct control of many of the actors and initiatives that we have explored in this book.

Last but not least, the examples throughout this book of governance in practice suggest it takes on a much broader range of forms in reality than many theoretical accounts allow for. We have clearly observed, as others have done in other areas, a shift from government to governance in which more actors are involved in processes of governing.[13] The tools of governance have not only broadened from law and regulation to voluntary standards, codes, and partnerships, but also to day-to-day supply chain management within firms whose decisions often dwarf those made in the traditional arenas of climate governance in terms of their impact on emissions of GHG. While the tremendous diversity and dynamism of climate governance generates huge challenges of coordination, accountability and effectiveness, a fact which in itself makes it hard to get a handle on how effective action is and who is benefiting, the plurality of sites of action could also be a positive thing as actors move between arenas trying to advance action in the fastest and most effective way they can, working with whom they need to, wherever that happens to be. We surely have to hope for all our sakes that the forms of climate governance we are now busy constructing are up to the scale of the challenge we are faced with, and can deliver change within the time we have available to us to prevent the very worst scenarios of uncontrolled climate change.

Notes

Foreword

1 See, for example, Christopher Booker, "2008 Was the Year Man-made Global Warming Was Disproved," *Daily Telegraph*, 27 December 2008.
2 IPCC, *Climate Change 2007: Synthesis Report* (Geneva, Switzerland: IPCC, 2007), 2–5.
3 See Michael G. Schechter, *United Nations Global Conferences* (London: Routledge, 2005).
4 See Lorraine Elliott, "Global Environmental Governance," in *Global Governance: Critical Perspectives,* eds Rorden Wilkinson and Steve Hughes (London: Routledge, 2002).
5 Most notably Elizabeth R. DeSombre, *Global Environmental Institutions* (London: Routledge, 2006).

Introduction

1 Greenpeace International, *The Oil Industry and Climate Change: A Greenpeace Briefing* (Amsterdam, Netherlands: Greenpeace International, 1998).
2 Michele Betsill and Harriet Bulkeley, "Transnational Networks and Global Environmental Governance: The Cities for Climate Protection Program," *International Studies Quarterly* 48, no. 2 (2004): 471–93; John Vogler, "Taking Institutions Seriously: How Regimes Can Be Relevant to Multilevel Environmental Governance," *Global Environmental Politics* 3, no. 2 (2003): 25–39; Neil Adger and Andrew Jordan, eds, *Governing Sustainability* (Cambridge: Cambridge University Press, 2009).
3 Geoffrey Heal, "New Strategies for the Provision of Global Public Goods: Learning from International Environmental Challenges," in *Global Public Goods: International Cooperation in the Twenty-first Century*, eds Inge Kaul, Isabelle Grunberg, and Marc Stern (Oxford: Oxford University Press, 1999), 240–264.
4 Adrian Smith, "Energy Governance: The Challenges of Sustainability," in *Energy for the Future: A New Agenda*, eds Ivan Scrase and Godron MacKeron (Basingstoke, UK: Palgrave Macmillan, 2009), 54–76.
5 Peter Newell and Matthew Paterson, "A Climate for Business: Global Warming, the State and Capital," *Review of International Political Economy* 5, no. 4 (1998): 679–704.

6 Andrew Hurrell, *International Politics of the Environment* (Oxford: Clarendon Press, 1992), 1.
7 Oran Young, "Rights, Rules and Resources in World Affairs," in *Global Governance: Drawing Insights from the Environmental Experience*, ed. Oran Young (Cambridge, Mass.: MIT Press, 1997), 5–6; see also Oran Young, *Global Governance: Learning Lessons from the Environmental Experience* (Cambridge, Mass.: MIT Press, 1998); eds Peter Haas, Robert Keohane, and Marc Levy, *Institutions for the Earth: Sources of Effective Environmental Protection* (Cambridge, Mass.: MIT Press, 1993).
8 Peter Newell, *Climate for Change: Non-State Actors and the Global Politics of the Greenhouse* (Cambridge: Cambridge University Press, 2000), 23–24.
9 Chukwumerije Okereke and Harriet Bulkeley, "Conceptualizing Climate Change Governance Beyond the International Regime: A Review of Four Theoretical Approaches," *Tyndall Working Paper* 112 (2007): 6.
10 Susan Strange, "Cave! Hic dragones: A Critique of Regime Analysis," *International Organization* 36, no. 2 (1983): 488–90.
11 Matthew Paterson, *Global Warming, Global Politics* (London: Routledge, 1996).
12 Susan Strange, "Cave! Hic dragones: A Critique of Regime Analysis," in *International Regimes*, eds Stephen Krasner (Ithaca, N.Y.: Cornell University Press, 1983), 337–354.
13 Karen Litfin, *Ozone Discourses: Science and Politics in Global Environmental Cooperation* (New York: Columbia University Press, 1994).
14 Okereke and Bulkeley, "Conceptualizing Climate Change Governance Beyond the International Regime: A Review of Four Theoretical Approaches," 6.
15 Harriet Bulkeley and Michele Betsill, *Cities and Climate Change: Urban Sustainability and Global Environmental Governance* (London: Routledge, 2003).
16 Newell, *Climate for Change: Non-State Actors and the Global Politics of the Greenhouse*.
17 Peter Haas, *Saving the Mediterranean: The Politics of International Environmental Cooperation* (New York: Columbia University Press, 1990), 349.
18 Paterson, *Global Warming, Global Politics*.
19 Paterson, *Global Warming, Global Politics*, 151.
20 Sheila Jasanoff and Brian Wynne, "Science and Decision-Making," in *Human Choice and Climate Change I: The Societal Framework*, eds Steve Rayner and Elizabeth Malone (Columbus, Ohio: Battelle Press, 1998), 1–87.
21 Matthew Auer, "Who Participates in Global Environmental Governance? Partial Answers from International Relations Theory," *Policy Sciences* 33, no. 2 (2000): 155–80; Bulkeley and Betsill, *Cities and Climate Change: Urban Sustainability and Global Environmental Governance*; Harriet Bulkeley, "Reconfiguring Environmental Governance: Towards a Politics of Scales and Networks," *Political Geography* 24, no. 8 (2005): 875–902.
22 Newell, *Climate for Change: Non-State Actors and the Global Politics of the Greenhouse*.
23 Peter Newell, "The Political Economy of Global Environmental Governance," *Review of International Studies* 34 (July 2008): 507–29.
24 Newell and Paterson, "A Climate for Business," 679–704; Newell, "The Political Economy of Global Environmental Governance," 507–29.
25 Bulkeley, "Reconfiguring Environmental Governance," 875–902.

26 Ole Jacob Sending and Iver B. Neumann, "Governance to Governmentality: Analyzing NGOs, States, and Power," *International Studies Quarterly* 50, no. 3 (2006): 651–72.

27 John Allen, "Powerful City Networks: More than Connections, Less than Domination and Control," *Urban Networks and Network Theory* (2008), www.lboro.ac.uk/gawc/rb/rb270.html

28 Thomas Risse, "Global Governance and Communicative Action," *Government and Opposition* 39, no. 2 (2004): 289.

29 James Rosenau, "Change, Complexity and Governance in a Globalizing Space," in *Debating Governance*, ed. Jon Pierre (Oxford: Oxford University Press, 2000), 172.

30 James Rosenau, "Governance, Order, and Change in World Politics," in *Governance Without Government: Order and Change in World Politics*, eds James Rosenau and Ernst Czempiel (Cambridge: Cambridge University Press, 1992), 6.

31 Klaus Dingwerth and Philipp Pattberg, "Global Governance as a Perspective on World Politics," *Global Governance* 12, no. 2 (2006): 185–203; Sverker Jagers and Johannes Stripple, "Climate Governance Beyond the State," *Global Governance* 9, no. 3 (2003): 385–99; Okereke and Bulkeley, "Conceptualizing Climate Change Governance Beyond the International Regime: A Review of Four Theoretical Approaches"; Matthew Paterson, David Humphreys, and Lloyd Pettiford, "Conceptualizing Global Environmental Governance: From Interstate Regimes to Counter-Hegemonic Struggles," *Global Environmental Politics* 3, no. 2 (2003): 1–8.

32 These are ideal types and there will of course be work which does not fit neatly into each category, while other authors may write from both perspectives. Note that we do not include here normative approaches to "global governance" associated with the drive by international organizations, nation-states and non-governmental bodies to promote "good" governance, though we acknowledge that such discourses share some liberal assumptions in common with academic approaches.

33 Sverker Jagers and Johannes Stripple, "Climate Governance Beyond the State," *Global Governance* 9, no. 3 (2003): 385.

34 Michele Betsill and Elisabeth Corell, eds, *NGO Diplomacy: The Influence of Nongovernmental Organizations in International Environmental Negotiations* (Cambridge, Mass.: MIT Press, 2008); Newell, *Climate for Change: Non-State Actors and the Global Politics of the Greenhouse*; Bas Arts, *The Political Influence of Global NGOs: Case Studies on the Climate and Biodiversity Conventions* (Utrecht, Netherlands: International Books, 1998).

35 Lars H. Gulbrandsen and Steinar Andresen, "NGO Influence in the Implementation of the Kyoto Protocol: Compliance, Flexibility Mechanisms, and Sinks," *Global Environmental Politics* 4, no. 4 (2004): 56.

36 Betsill and Corell, *NGO Diplomacy: The Influence of Nongovernmental Organizations in International Environmental Negotiations*, 3.

37 Michele Betsill, "Transnational Actors in International Environmental Politics," in *Palgrave Advances in International Environmental Politics*, eds Michele Betsill, Kathryn Hochstetler, and Dimitris Stevis (Basingstoke, UK: Palgrave Macmillan, 2006), 191.

38 Dingwerth and Pattberg, "Global Governance as a Perspective on World Politics," 197.

39 Jagers and Stripple, "Climate Governance Beyond the State," 385.
40 Paul Wapner, *Environmental Activism and World Civic Politics* (Albany, N.Y.: State University of New York Press, 1996); Ronnie Lipschutz, "From Place to Planet: Local Knowledge and Global Environmental Governance," *Global Governance* 3, no. 1 (1997): 83–102.
41 Bulkeley and Betsill, *Cities and Climate Change: Urban Sustainability and Global Environmental Governance*; Henrik Selin and Stacy D. VanDeveer, "Canadian-US Environmental Cooperation: Climate Change Networks and Regional Action," *American Review of Canadian Studies* 35, no. 2 (2005): 353–78.
42 Karin Bäckstrand, "Accountability of Networked Climate Governance: The Rise of Transnational Climate Partnerships," *Global Environmental Politics* 8, no. 3 (2008): 74–102; Philipp Pattberg and Johannes Stripple, "Beyond the Public and Private Divide: Remapping Transnational Climate Governance in the 21st Century," *International Environmental Agreements: Politics, Law and Economics* 8, no. 4 (2008): 367–388; Thorsten Benner, Wolfgang H. Reinicke, and Jan Martin Witte, "Multisectoral Networks in Global Governance: Towards a Pluralistic System of Accountability," *Government and Opposition* 39, no. 2 (2004): 191–210.
43 Benjamin Cashore, "Legitimacy and the Privatization of Environmental Governance: How Non-State Market-Driven Governance Systems Gain Rule-Making Authority," *Governance* 15, no. 4 (2004): 503–29; Jennifer Clapp, "The Privatisation of Global Environmental Governance: ISO 14001 and the Developing World," *Global Governance* 4, no. 3 (1998): 295–316; Klaus Dingwerth, "North–South Parity in Global Governance: The Affirmative Procedures of the Forest Stewardship Council," *Global Governance* 14, no. 1 (2008): 53–71; Simone Pulver, "Importing Environmentalism: Explaining Petroleos Mexicanos' Proactive Climate Policy," *Studies in Comparative International Development* 42, no. 3/4 (2007): 233–55.
44 Paul Wapner, "Governance in Global Civil Society," in *Global Governance: Drawing Insight from Environmental Experience*, ed. Oran Young (Cambridge, Mass.: MIT Press, 1997), 67.
45 David Levy and Peter Newell, eds, *The Business of Global Environmental Governance* (Cambridge, Mass.: MIT Press, 2005); Matthew Paterson, *Understanding Global Environmental Politics: Domination, Accumulation and Resistance* (Houndmills, UK: Macmillan, 2000); Julian Saurin, "International Relations, Social Ecology and Globalization of Environmental Change," in *The Environment and International Relations*, eds John Vogler and Mark Imber (London: Routledge, 1996), 77–99; Julian Saurin, "Global Environmental Crisis as 'Disaster Triumphant': The Private Capture of Public Goods," *Environmental Politics* 10, no. 4 (2001): 80.
46 Bob Jessop, "Liberalism, Neoliberalism, and Urban Governance: A State-Theoretical Perspective," *Antipode* 34, no. 3 (2002): 463.
47 Chukwumerije Okereke, Harriet Bulkeley, and Heike Schroeder, "Conceptualizing Climate Governance Beyond the International Regime," *Global Environmental Politics* 9, no. 1 (2009): 58–78.
48 David Levy and Peter Newell, "Business Strategy and International Environmental Governance: Towards a Neo-Gramscian Synthesis," *Global Environmental Politics* 2, no. 4 (2002): 84–101; Levy and Newell, *The Business of Global Environmental Governance*.

49 Peter Newell, "The Political Economy of Global Environmental Governance," *Review of International Studies* 34, no. 3 (2008): 507–29.

50 Michael Ekers and Alex Loftus, "The Power of Water: Developing Dialogues Between Foucault and Gramsci," *Environment and Planning D: Society and Space* 26, no. 4 (2008): 703.

51 Ole J. Sending and Iver B. Neumann, "Governance to Governmentality: Analyzing NGOs, States, and Power," *International Studies Quarterly* 50, no. 3 (2006): 653.

52 Sending and Neumann, "Governance to Governmentality: Analyzing NGOs, States, and Power," 657.

1 Governing climate change: a brief history

1 Joanna Depledge, *The Organization of the Global Negotiations: Constructing the Climate Regime* (London: Earthscan, 2005).

2 Matthew Paterson and Michael Grubb, "The International Politics of Climate Change," *International Affairs* 68, no. 2 (1992): 293–310.

3 Peter Newell, *Climate for Change: Non-State Actors and the Global Politics of the Greenhouse* (Cambridge: Cambridge University Press, 2000).

4 Bas Arts, *The Political Influence of Global NGOs: Case Studies on the Climate and Biodiversity Conventions* (Utrecht, Netherlands: International Books, 1998); Peter Newell, *Climate for Change*.

5 Michele Betsill and Elisabeth Corell, eds, *NGO Diplomacy: The Influence of Nongovernmental Organizations in International Environmental Negotiations* (Cambridge, Mass.: MIT Press, 2008), 3.

6 United Nations, *United Nations Framework Convention on Climate Change*, Bonn, Germany: UNFCCC Secretariat, 1992, http://unfccc.int/resource/docs/convkp/conveng.pdf

7 Michael Grubb with Christiaan Vrolijk and Duncan Brack, *The Kyoto Protocol: A Guide and Assessment* (London: Earthscan and the Royal Institute of International Affairs, 1999).

8 Michael Grubb and Farhana Yamin, "Climatic Collapse at The Hague: What Happened, Why, and Where Do We Go from Here?" *International Affairs* 77, no. 2 (2001): 261–76.

9 Chukwumerije Okereke, Phillip Mann, and Andy Newsham, "Assessment of Key Negotiating Issues at Nairobi Climate COP/MOP and what It Means for the Future of the Climate Regime," *Tyndall Centre Working Paper* 106 (2007).

10 Benito Muller, "Bali 2007: On the Road Again! Impressions from the Thirteenth UN Climate Change Conference," www.oxfordclimatepolicy.org/publications/Bali2007Final.pdf; Jennifer Morgan, "Towards a New Global Climate Deal: An Analysis of the Agreements and Politics of the Bali Negotiations. E3G," www.e3g.org/images/uploads/Bali_Analysis_Morgan_080120.pdf

11 Bert Bolin, "Scientific Assessment of Climate Change," in *International Politics of Climate Change: Key Issues and Actors*, ed. Gunnar Fermann (Oslo, Norway: Scandinavian University Press, 1997), 99.

12 Michel Foucault, *Power/Knowledge: Selected Interviews and Other Writings 1972–77* (New York: Pantheon books, 1980), 27.

13 Bolin, "Scientific Assessment of Climate Change," 102.

14 Peter Haas, *Saving the Mediterranean: The Politics of International Environmental Cooperation* (New York: Columbia University Press, 1990).

15 Karen Litfin, *Ozone Discourses* (New York: Columbia University Press, 1994).
16 Sonia Boehmer-Christiansen, "Global Climate Protection Policy: The Limits of Scientific Advice—Part 1," *Global Environmental Change* 4, no. 2 (1992): 140–59.
17 Among the most prominent "climate skeptics" are Fred Singer (George Mason University) and Richard Linzen (MIT).
18 "US Reportedly Seeking to Sink Watson as IPCC Head," www.unwire.org/unwire/20020402/25249_story.asp
19 Article 2 of the UNFCCC (1992): http://unfccc.int/resource/docs/convkp/conveng.pdf
20 *ECO* NGO Newsletter, issue 7, Geneva, 1990.
21 Adil Najam, Saleemul Huq, and Youba Sokona, "Climate Negotiations Beyond Kyoto: Developing Countries' Concerns and Interests," *Climate Policy* 3, no. 3 (2003): 221–31; Michele Williams, "The Third World and Global Environmental Negotiations: Interests, Institutions and Ideas," *Global Environmental Politics* 5, no. 3 (2005): 48–49.
22 Anil Argwal and Sunita Narain, *"Global Warming in an Unequal World: A Case of Environmental Colonialism,"* (New Delhi, India: Centre for Science and the Environment/WRI, 1991).
23 Article 3 of the UNFCCC (1992): http://unfccc.int/resource/docs/convkp/conveng.pdf
24 Zoe Young, *A New Green Order? The World Bank and the Politics of the Global Environment Facility* (London: Pluto Press, 2002).
25 Sjur Kasa, Anne Therese Gullberg, and Gørild Heggelund, "The Group of 77 in the International Climate Negotiations: Recent Developments and Future Directions," *International Environmental Agreements: Politics, Law and Economics* 8, no. 2 (2008): 113–27.
26 Scott Barrett, "Montreal Versus Kyoto: International Cooperation on the Global Environment," in *Global Public Goods*, eds Inge Kaul, Isabelle Grunberg, and Marc Stern (New York: Oxford University Press, 1999), 192–220.
27 Jan-Peter Voss, "Innovation Processes in Governance: the Development of 'Emissions Trading' as a New Policy Instrument," *Science and Public Policy* 34, no. 5 (2007): 329–43; Jon Birger Skjærseth and Jørgen Wettestad, *EU Emissions Trading: Initiation, Decision-Making and Implementation* (Aldershot, UK: Ashgate, 2008).
28 Miranda Schreurs and Elizabeth Economy, eds, *The Internationalization of Environmental Protection* (Cambridge: Cambridge University Press, 1997); Peter Newell "The Political Economy of Global Environmental Governance," *Review of International Studies* 34 (July 2008): 507–29.
29 Newell, *Climate for Change.*
30 David Levy and Peter Newell, eds, *The Business of Global Environmental Governance* (Cambridge, Mass.: MIT Press, 2005); Chukwumerije Okereke, Harriet Bulkeley, and Heike Schroeder, "Conceptualizing Climate Governance Beyond the International Regime," *Global Environmental Politics* 9, no. 1 (2009): 58–78.

2 Governance for whom? Equity, justice, and the politics of sustainable development

1 Peter Newell, "Climate for Change: Civil Society and the Politics of Global Warming," in *Global Civil Society Yearbook,* eds Marlies Glasius, Mary Kaldor and Helmut Anheier (London: Sage, 2005), 90–119.

2 Timmons Roberts and Bradley C. Parks, *A Climate of Injustice: Global Inequality, North–South Politics, and Climate Policy* (Cambridge, Mass.: MIT Press, 2007); Chukwumerije Okereke, *Global Justice and Neoliberal Environmental Governance: Ethics, Sustainable Development and International Co-operation* (London: Routledge, 2008).

3 Heidi Bachram, "Climate Fraud and Carbon Colonialism: The New Trade in Greenhouse Gases," *Capitalism, Nature, Socialism* 15, no. 4 (2004): 10–12.

4 Bradley C. Parks and Timmons Roberts, "Globalization, Vulnerability to Climate Change and Perceived Injustice," *Society and Natural Resources* 19, no. 4 (2006): 342; Tearfund, *Dried Up, Drowned Out: Voices from the Developing World on a Changing Climate* (London: Tearfund, 2005); Peter Newell, "Race, Class and the Global Politics of Environmental Inequality," *Global Environmental Politics* 5, no. 3 (2005): 70–94.

5 "The Case for Climate Security," lecture by the Foreign Secretary, the Rt. Hon. Margaret Beckett MP, Royal United Services Institute, 10 May 2007, www.rusi.org/events/past/ref:E464343E93D15A/info:public/infoID: E4643430E3E85A/

6 Quoted in Peter Newell, "Climate Change, Human Rights and Corporate Accountability," in *Climate Change and Human Rights*, ed. Stephen Humphrey (Cambridge: Cambridge University Press, 2009, 126–159).

7 Heidi Bachram, "Climate Fraud and Carbon Colonialism: The New Trade in Greenhouse Gases," 10–12.

8 Raúl Estrada, "First Approaches and Unanswered Questions," in *Issues and Options: The Clean Development Mechanism* (New York: UNDP, 1998), 23–29.

9 Roberts and Parks, *A Climate of Injustice: Global Inequality, North–South Politics, and Climate Policy*.

10 Sjur Kasa, Anne Therese Gullberg, and Gørild Heggelund, "The Group of 77 in the International Climate Negotiations: Recent Developments and Future Directions," *International Environmental Agreements: Politics, Law and Economics* 8, no. 2 (2008): 113–27.

11 Peter Newell, Nicky Jenner, and Lucy Baker, "Governing Clean Development: A Framework for Analysis," *The Governance of Clean Development Working Paper Series*, no. 1 (Norwich, UK: UEA, 2009), www.clean-development.com

12 Ian Tellam, ed., *Fuel for Change: World Bank Energy Policy-Rhetoric and Reality* (London: Zed Books, 2000), 33.

13 World Bank, *Carbon Funds and Facilities*, http://go.worldbank.org/51X7C H8VN0

14 World Resources Institute, *Correcting the World's Greatest Market Failure: Climate Change and Multilateral Development Banks* (2008), www.wri.org/publication/correcting-the-worlds-greatest-market-failure

15 According to the World Bank, "new renewable energy" applies to energy from biomass, solar, wind, geothermal and small hydro (under 10MW) systems.

16 Practical Action, *Energy to reduce poverty* (2007), http://practicalaction.org/docs/advocacy/energy-to-reduce-poverty_g8.pdf

17 Davidson, Ogunlade, Kirsten Halsnæs, Saleemul Huq, Marcel Kok, Bert Metz, et al., "The Development and Climate Nexus: The Case of Sub-Saharan Africa," *Climate Policy* 3, no. 1 (2003): 97–113.

18 UNFCCC, *Investment and Financial Flows to Address Climate Change* (Bonn, Germany: UNFCCC, 2007).

19 UNFCCC, *Investment and Financial Flows to Address Climate Change*, 26.
20 UNFCCC, *Investment and Financial Flows to Address Climate Change*, Executive summary.
21 Emily Boyd, Nathan Hultman, Timmons Roberts, Esteve Corbera, Johannes Ebeling, et al., "The Clean Development Mechanism: An Assessment of Current Practice and Future Approaches for Policy," *Tyndall Centre Working Paper* 114, www.tyndall.ac.uk/publications/working_papers/twp114_summary.shtml; Karen. H. Olsen, "The Clean Development Mechanism's Contribution to Sustainable Development: A Review of the Literature," *Climate Change* 84, no. 1 (2007): 59–73.
22 Newell, Jenner and Baker, "Governing Clean Development: A Framework for Analysis."
23 Newell, Jenner and Baker, "Governing Clean Development: A Framework for Analysis."
24 World Trade Organization, *Activities of the WTO and the Challenge of Climate Change*, www.wto.org
25 World Bank, *International Trade and Climate Change: Economic, Legal and Institutional Perspectives* (Washington, DC: World Bank, 2007).
26 Neil Adger, "Scales of Governance and Environmental Justice for Adaptation and Mitigation of Climate Change," *Journal of International Development* 13, no. 7 (2001): 921–31.
27 World Bank Group, *Poverty and Climate Change: Reducing the Vulnerability of the Poor Through Adaptation* (Washington, DC: World Bank Group, 2003).
28 Parks and Roberts, "Globalization, Vulnerability to Climate Change and Perceived Injustice," 342.
29 Parks and Roberts "Globalization, Vulnerability to Climate Change and Perceived Injustice," 345.
30 Jouni Paavola and Neil Adger, "Justice and Adaptation to Climate Change," *Tyndall Centre Working Paper* no. 23, October 2002.
31 Bachram, "Climate Fraud and Carbon Colonialism: The New Trade in Greenhouse Gases," 10–12.
32 Jutta Kill, "Land Grab in Uganda in Preparation for CDM Sinks Project," *World Rainforest Movement Bulletin* no. 74, September 2003.
33 "Sinks Watch," www.sinkswatch.org
34 Kevin Watkins, editor of the UNDP *Human Development Report*, at its launch in Brazil, 27 November. Quoted in Larry Elliot and Ashley Seager, "Cut Carbon By Up to Third to Save Poor, UN Tells West," *Guardian*, 28 November 2007, www.guardian.co.uk/environment/2007/nov/28/climatechange
35 Peter Newell, "Civil Society, Corporate Accountability and the Politics of Climate Change," *Global Environmental Politics* 8, no. 3 (2008): 124–55.
36 Earth Justice, "Petition to the Inter-American Commission on Human Rights Seeking Relief From Violations Resulting From Global Warming Caused By Acts and Omissions of the United States," www.earthjustice.org/library/legal_docs/summary-of-inuit-petition-to-inter-american-council-on-human-rights.pdf
37 Earth Justice, ibid.
38 "Corporate Watch," (2007), www.corpwatch.org/article.php?id=979
39 Newell, "Climate for Change: Civil Society and the Politics of Global Warming."

40 New Economics Foundation, *Collision Course: Free Trade's Free Ride on the Global Climate* (London: New Economics Foundation, 2003).

41 Christopher Swan, "Zoellick Fossil Fuel Campaign Belied by World Bank's Tata Loan," *Bloomberg.com*, 10 August 2008.

42 Oliver Tickell and Nicola Hildyard, "Green Dollars, Green Menace," *The Ecologist* 22, no. 3 (1992): 82–83.

3 Between global and local: governing climate change transnationally

1 Throughout the chapter, the terms "network" and "arrangement" are used interchangeably to describe transnational collaboration between various types of actor. Not all of these phenomena could be classified as "networks" in organizational terms, and nor should the term imply that power relations among those participating are necessarily absent or that each is equally active or vital to the network.

2 Robert Keohane and Joseph Nye, *Transnational Relations and World Politics* (Cambridge, Mass.: Harvard University Press, 1971).

3 Liliana Andonova and Marc Levy, "Franchising Global Governance: Making Sense of the Johannesburg Type II Partnerships," in *Yearbook of International Co-operation on Environment and Development 2003/2004* (London: Earthscan Publications, 2003); Ben Cashore, Graeme Auld, and Deanna Newsom, *Governing through Markets: Forest Certification and the Emergency of Non-State Authority* (New Haven, Conn.: Yale University Press, 2004); Jennifer Clapp, "The Privatisation of Global Environmental Governance: ISO 14001 and the Developing World," *Global Governance* 4, no. 3 (1998): 295–316; Pieter Glasbergen, Frank Biermann, and Arthur P. J. Mol, *Partnerships, Governance and Sustainable Development: Reflections on Theory and Practice* (Cheltenham, UK: Edward Elgar, 2007); Peter M. Haas, "Do Regimes Matter? Epistemic Committees and Mediterranean Pollution Control," *International Organization* 43, no. 3 (1989): 377–403; Margaret E. Keck and Kathryn Sikkink, *Activists Beyond Borders: Advocacy Networks in International Politics* (Ithaca, N.Y.: Cornell University Press, 1998); Ann-Marie Slaughter, *A New World Order* (Princeton, N.J.: Princeton University Press, 2004).

4 Once known as the International Council for Local Environmental Initiatives, ICLEI now goes by its acronym and the title Local Governments for Sustainability.

5 Karin Backstränd, "Accountability of Networked Climate Governance: The Rise of Transnational Climate Partnerships," *Global Environmental Politics* 8, no. 3 (2008): 75.

6 Liliana Andonova, Michele Betsill, and Harriet Bulkeley, "Transnational Climate Governance," *Global Environmental Politics* 9, no. 2 (2009): 52–73.

7 Andonova et al., "Transnational Climate Governance"; Philipp Pattberg and Johannes Stripple, "Beyond the Public and Private Divide: Remapping Transnational Climate Governance in the 21st Century," *International Environmental Agreements: Politics, Law and Economics* 8, no. 4 (2008): 367–88.

8 Andonova et al., "Transnational Climate Governance."

9 Thorsten Benner, Wolfgang Reinicke, and Jan Witte, "Multisectoral Networks in Global Governance: Towards a Pluralistic System of Accountability," *Government and Opposition* 39, no. 2 (2004): 191–210.

10 Henrik Selin and Stacy VanDeveer, "Canadian-US Environmental Cooperation: Climate Change Networks and Regional Action," *American Review of
Canadian Studies* (Summer 2005): 253–78.

11 Harriet Bulkeley and Michele Betsill, *Cities and Climate Change: Urban
Sustainability and Global Environmental Governance* (London: Routledge,
2003); Kristine Kern and Harriet Bulkeley, "Cities, Europeanization and
Multi-level Governance: Governing Climate Change through Transnational
Municipal Networks," *JCMS* 47, no. 2 (2009): 309–32.

12 Kern and Bulkeley, "Cities, Europeanization and Multi-level Governance."

13 Bulkeley and Betsill, *Cities and Climate Change.*

14 ICLEI, *Programs: CCP Participants,* www.iclei.org/index.php?id=809

15 ICLEI, *Members,* www.iclei.org/index.php?id=global-members

16 Kern and Bulkeley, "Cities, Europeanization and Multi-level Governance."

17 City of Seattle Homepage, "Who is Involved—US Mayors Climate Protection
Agreement," www.seattle.gov/mayor/climate/default.htm#who

18 Climate Mayors: Australasian Mayors Council for Climate Protection,
"Hot Topic: 'Step Up' to CCP Partner," www.iclei.org/index.php?id=6226;
ICLEI, "About ICLEI: Advocacy," www.iclei.org/index.php?id=2260

19 "C40 Cities, an Introduction," www.c40cities.org/

20 Clinton Foundation, www.clintonfoundation.org/051607-nr-cf-pr-cci-president-
clinton-announces-landmark-program-to-reduce-energy-use-in-buildings-world
wide.htm

21 Kern and Bulkeley, "Cities, Europeanization and Multi-level Governance,"
319–20.

22 Energie-Cities, "Info Bulletin 33," 2007.

23 Harriet Bulkeley, "Urban Sustainability: Learning from Best Practice?"
Environment and Planning A 38, no. 5 (2006): 235–54.

24 Bulkeley and Betsill, *Cities and Climate Change.*

25 Michele Betsill and Harriet Bulkeley, "Transnational Networks and Global
Environmental Governance: The Cities for Climate Protection Program,"
International Studies Quarterly 48, no. 2 (2004): 471–93; Bulkeley, "Urban
Sustainability: Learning from Best Practice?"

26 Climate Alliance, *Activity Report* (Frankfurt/Main: Climate Alliance, 2007).

27 Harriet Bulkeley, "Down to Earth: Local Government and Greenhouse
Policy in Australia," *Australian Geographer* 31, no. 3 (2000): 289–308.

28 Kern and Bulkeley, "Cities, Europeanization and Multi-level Governance."

29 Bulkeley and Betsill, *Cities and Climate Change.*

30 Andonova and Levy, "Franchising Global Governance: Making Sense of
the Johannesburg Type II Partnerships"; Backstränd, "Accountability of
Networked Climate Governance: The Rise of Transnational Climate Part-
nerships"; Benner et al., "Multisectoral Networks in Global Governance";
Frank Biermann, Arthur P. J. Mol, and Pieter Glasbergen, "Conclusion:
Partnerships for Sustainability—Reflections on a Future Research Agenda,"
in *Partnerships, Governance and Sustainable Development, Reflections on
Theory and Practice,* eds Frank Biermann, Arthur P. J. Mol, and Pieter
Glasbergen (Cheltenham: Edward Elgar, 2007), 288–300; Pattberg and
Stripple, "Beyond the Public and Private Divide."

31 Backstränd, "Accountability of Networked Climate Governance: The Rise
of Transnational Climate Partnerships," 77; Pattberg and Stripple, "Beyond
the Public and Private Divide."

32 Backstränd, "Accountability of Networked Climate Governance," 89.
33 REEEP, "Welcome to REEEP," www.reeep.org/31/home.htm
34 Backstränd, "Accountability of Networked Climate Governance," 90–91.
35 Charlotte Streck, "New Partnerships in Global Environmental Policy: The Clean Development Mechanism," *Journal of Environment and Development* 13, no. 3 (2004): 295.
36 Streck, "New Partnerships in Global Environmental Policy: The Clean Development Mechanism," 297.
37 CDM Watch, "Market Failure: Why the Clean Development Mechanism Won't Promote Clean Development," www.sinkswatch.org/pubs/Market%20failure.pdf
38 The Climate Group, "States and Regions," www.theclimategroup.org/what_we_do/states_and_regions
39 The Climate Group, "Voluntary Carbon Standard," www.theclimategroup.org/what_we_do/vcs
40 John G. Ruggie, "Reconstituting the Global Public Domain—Issues, Actors and Practices," *European Journal of International Relations* 10, no. 4 (2004): 504.
41 Ruggie, "Reconstituting the Global Public Domain—Issues, Actors and Practices," 504.
42 Cashore et al., *Governing through Markets: Forest Certification and the Emergency of Non-State Authority*; Clapp, "The Privatisation of Global Environmental Governance: ISO 14001 and the Developing World"; Klaus Dingwerth, "North–South Parity in Global Governance: The Affirmative Procedures of the Forest Stewardship Council," *Global Governance: A Review of Multilateralism and International Organizations* 14, no. 1 (2008): 53–73; Pattberg and Stripple, "Beyond the Public and Private Divide."
43 Dingwerth, "North–South Parity in Global Governance," 608.
44 Backstränd, "Accountability of Networked Climate Governance," 95.
45 Ingrid J. Visseren-Hamakers and Bas Arts, "Interaction Management by Partnerships: The Case of Biodiversity and Climate Change Governance System Interaction," paper for the Private Environmental Regimes Conference, 2008.
46 CCB Standards, "CCB Projects," www.climate-standards.org/projects/index.html
47 Clapp, "The Privatization of Global Environmental Governance."
48 Ruggie, "Reconstituting the Global Public Domain—Issues, Actors and Practices," 504.
49 ICLEI Oceania, *Local Government Action on Climate Change: CCP Australia Measures Evaluation Report 2008* (Melbourne, Australia: ICLEI Oceania and Australian Government, 2008), www.iclei.org/index.php?id=ccp-measures08
50 Peter Newell, "Civil Society, Corporate Accountability and the Politics of Climate Change," *Global Environmental Politics* 8, no. 3 (2008): 124–55.

4 Community and the governing of climate change

1 Ken Conca, "Greening the UN: Environmental Organizations and the UN System," *Third World Quarterly* 16, no. 3 (1995): 103–19; Thomas Princen and Matthias Finger, *Environmental NGOs in World Politics* (London:

Routledge, 1994); Paul Wapner, *Environmental Activism and World Civic Politics* (Albany: State University of New York Press, 1996).

2 Matthew R. Auer, "Who Participates in Global Environmental Governance? Partial Answers From International Relations Theory," *Policy Sciences* 33, no. 2 (2000): 163.

3 Susan Owens, "Engaging the Public: Information and Deliberation in Environmental Policy," *Environment and Planning A* 32, no. 7 (2000): 1141–48.

4 Brian Wynne, "May the Sheep Graze Safely? A Reflexive View of the Expert-Lay Divide" in *Risk, Environment and Modernity: Towards a New Ecology*, ed. Scott Lash, Bronislaw Szerszynski, and Brian Wynne (London: Sage, 1996), 44–83; Simon Joss, ed., "Public Participation in Science and Technology," special issue, *Science and Public Policy* 26, no. 5 (1999); Brian Wynne, "Creating Public Alienation: Expert Cultures of Risk and Ethics on GMOs," *Science as Culture* 10, no. 4 (2001): 445–81.

5 Irene Lorenzoni, Sophie Nicholson-Cole, and Lorraine Whitmarsh, "Barriers Perceived to Engaging with Climate Change Among the UK Public and Their Policy Implications," *Global Environmental Change: Human and Policy Dimensions* 17, no. 3–4 (2007): 445–59.

6 Steve Hinchliffe, "Helping the Earth Begins at Home—The Social Construction of Socio-Environmental Responsibilities," *Global Environmental Change—Human and Policy Dimensions* 6, no. 1 (1996): 53–62.

7 Hinchliffe, "Helping the Earth Begins at Home."

8 Carlo Jaeger, Gregor Durrenberger, Hans Kastenholz, and Bernhard Truffer, "Determinants of Environmental Action With Regard to Climatic-Change," *Climatic Change* 23, no. 3 (1993): 193–211.

9 Willet Kempton, James Boster, and Jennifer Hartley, *Environmental Values in American Culture* (Boston, Mass.: MIT Press, 1995).

10 Owens, "Engaging the Public"; Walter Baber and Robert V. Bartlett, *Deliberative Environmental Politics: Democracy and Ecological Rationality* (Cambridge, Mass.: MIT Press, 2005); Minu Hemmati, *Multi-Stakeholder Processes for Governance and Accountability* (London: Earthscan, 2002).

11 Judith Petts and Catherine Brooks, "Expert Conceptualizations of the Role of Lay Knowledge in Environmental Decision-Making: Challenges for Deliberative Democracy," *Environment and Planning A* 38, no. 6 (2006): 1047.

12 Steve Yearley, Steve Cinderby, John Forrester, Peter Bailey and Paul Rosen, "Participatory Modelling and the Local Governance of the Politics of UK Air Pollution: A Three-City Case Study," *Environmental Values* 12, no. 2 (2003): 247–62.

13 Owens, "Engaging the Public," 1144.

14 Roger Few, Kate Brown, and Emma Tomkins, "Public Participation and Climate Change Adaptation: Avoiding the Illusion of Inclusion," *Climate Policy* 7, no. 1 (2007): 46–59.

15 John Allen, "The Whereabouts of Power: Politics, Government and Space," *Geografiska Annaler* 86B, no. 1 (2004): 17–30.

16 Allen, "The Whereabouts of Power," 83.

17 Allen, "The Whereabouts of Power," 86.

18 Maarten van Aalst, Terry Cannon, and Ian Burton, "Community Level Adaptation to Climate Change: The Potential Role of Participatory Community Risk Assessment," *Global Environmental Change* 18, no. 1 (2008): 165–79.

19 Van Aalst et al., "Community Level Adaptation to Climate Change," 166.
20 Van Aalst et al., "Community Level Adaptation to Climate Change," 168.
21 Van Aalst et al., "Community Level Adaptation to Climate Change," 180.
22 Allen, "The Whereabouts of Power," 94.
23 Peter Miller and Nikolas Rose, "Governing Economic Life," *Economy and Society* 19, no. 1 (1990): 8; Jennifer A. Summerville, Barbara A. Adkins, and Gavin Kendall, "Community Participation, Rights, and Responsibilities: the Governmentality of Sustainable Development Policy in Australia," *Environment and Planning C: Government and Policy* 26, no. 4 (2008): 697.
24 Summerville et al., "Community Participation, Rights, and Responsibilities: the Governmentality of Sustainable Development Policy in Australia," 697.
25 Emily Boyd, Maria Gutierrez, and Manyu Chang, "Small-Scale Forest Carbon Projects: Adapting CDM to Low Income Communities," *Global Environmental Change* 17, no. 2 (2007): 250–59.
26 Gordon Walker, Sue Hunter, Patrick Devine-Wright, Bob Evans, and Helen Fay, "Harnessing Community Energies: Explaining and Evaluating Community-Based Localism in Renewable Energy Policy in the UK," *Global Environmental Politics* 7, no. 2 (2007): 64–82.
27 William M. Adams, *Green Development: Environment and Sustainability in the South* (London: Routledge, 2001).
28 Adams, *Green Development*.
29 Boyd et al., "Small-Scale Forest Carbon Projects," 251.
30 Boyd et al., "Small-Scale Forest Carbon Projects," 251.
31 Adam Bumpus and Diana Liverman, "Accumulation By Decarbonisation and the Governance of Carbon Offsets," *Economic Geography* 84, no. 2 (2008): 127–56; In 2007, 18 percent of projects in the voluntary market were made up of forestry projects. See Katherine Hamilton, Milo Sjardin, Thomas Marcello, and Gordon Xu, "Forging a Frontier: State of the Voluntary Carbon Markets," in *Ecosystem Marketplace and New Carbon Finance* (2008), www.ecosystemmarketplace.com/documents/cms_documents/2008_StateofVoluntaryCarbonMarket2.pdf
32 Boyd et al., "Small-Scale Forest Carbon Projects: Adapting CDM to Low Income Communities," 254.
33 Esteve Corbera, Carmen González Soberanis, and Katrina Brown, "Institutional Dimensions of Payments for Ecosystem Services: An Analysis of Mexico's Carbon Forestry Programme," *Ecological Economics* 68, no. 3 (2009): 743–61.
34 Corbera et al., "Institutional Dimensions of Payments for Ecosystem Services: An Analysis of Mexico's Carbon Forestry Programme."
35 Corbera et al., "Institutional Dimensions of Payments for Ecosystem Services," 751.
36 Corbera et al., "Institutional Dimensions of Payments for Ecosystem Services," 751.
37 Corbera, et al., "Institutional Dimensions of Payments for Ecosystem Services," 755.
38 Bumpus and Liverman, "Accumulation by Decarbonisation and the Governance of Carbon Offsets."
39 Bumpus and Liverman, "Accumulation by Decarbonisation and the Governance of Carbon Offsets."

40 Frances Seymour, "Forests, Climate Change and Human Rights," note prepared for the meeting of the International Council on Human Rights Policy (Geneva, Switzerland: 12–13 October 2007), 4.

41 Walker et al., "Harnessing Community Energies," 65.

42 Walker et al., "Harnessing Community Energies," 70.

43 Gordon Walker, "What are the Barriers and Incentives for Community-Owned Means of Energy Production and Use?" *Energy Policy* 36, no. 12 (2008): 4401.

44 Baywind Energy, homepage, www.baywind.co.uk

45 Walker, "What are the Barriers and Incentives for Community-Owned Means of Energy Production and Use?" 4401.

46 Gordon Walker and Patrick Devine-Wright, "Community Renewable Energy: What Should It Mean?" *Energy Policy* 36, no. 2 (2008): 498–99.

47 Walker et al., "Harnessing Community Energies," 71–73.

48 John Barry, Geraint Ellis, and Clive Robinson, "Cool Rationalities and Hot Air: A Rhetorical Approach to Understanding Debates on Renewable Energy," *Global Environmental Politics* 8, no. 2 (2008): 67–98.

49 Walker et al., "Harnessing Community Energies," 73.

50 Transition Towns Wiki, "What is a Transition Town (or village/city/forest/island)?" http://transitiontowns.org/TransitionNetwork/TransitionInitiative

51 Rob Hopkins and Peter Lipman, "The Transition Network Ltd: Who We Are And What We Do ... ," www.transitionnetwork.org/Strategy/Transition Network-WhoWeAreWhatWeDo.pdf; http://transitiontowns.org/Transition Network/TransitionNetwork

52 Kelvin Mason and Mark Whitehead, "Transition Urbanism and the Contested Politics of Spatial Practice," *Annual International Conference of the Royal Geographical Society with the Institute of British Geographers,* 2008.

53 Mason and Whitehead, "Transition Urbanism and the Contested Politics of Spatial Practice," 8.

54 Transition Town Totnes, *Food Group Homepage,* http://totnes.transitionnet work.org/food/home

55 Transition Town Brixton, *Food and Growing Group,* www.site.transitiontown brixton.org/index.php?option=com_content&view=article&id=117&Itemid=63

56 Transition Town Brixton, *Buildings and Energy Group* www.site.transitiontown brixton.org/index.php?option=com_content&view=article&id=135&Itemid=64

57 Sustainable Berea, *Home Energy and Consevation Project,* http://sustainable berea.org/home-enery-conservation-project

5 The private governance of climate change

1 Stephen Schmidheiny and the Business Council for Sustainable Development, *Changing Course* (Boston, Mass.: MIT Press, 1992), 43.

2 Greenpeace International, "The Oil Industry and Climate Change: A Greenpeace Briefing" (Amsterdam, Netherlands: Greenpeace International, 1998).

3 UNFCCC, "Investment and Financial Flows to Address Climate Change" (Bonn, Germany: UNFCCC, 2007).

4 International Chamber of Commerce, "Statement by the International Chamber of Commerce Before COP1," Berlin, Germany, 29 March 1995.

5 David Levy, "Business and the Evolution of the Climate Regime: The Dynamics of Corporate Strategies," in *The Business of Global Environmental*

Governance, eds David Levy and Peter Newell (Cambridge, Mass.: MIT Press, 2005), 73–105.

6 Cited in Peter Newell, *Climate for Change: Non-State Actors and the Global Politics of the Greenhouse* (Cambridge: Cambridge University Press, 2000), 100.

7 "Europe's Industries Play Dirty," *Economist*, 9 May 1992.

8 Peter Newell and Matthew Paterson, "A Climate for Business: Global Warming, the State and Capital," *Review of International Political Economy* 5, no. 4 (Winter 1998): 679–704.

9 Carl Deal, *The Greenpeace Guide to Anti-Environmental Organizations* (Berkeley, Calif.: Odonian Press, 1993).

10 Ross Gelbspan, *The Heat Is on: The Climate Crisis, the Cover-Up, the Prescription* (Reading, Mass.: Perseus Books Group, 1998).

11 Danny Hakim, "The Media Business: Ads are a Weapon in the Battle Over Higher Fuel Standards," *New York Times*, 12 March 2002, www.nytimes.com/2002/03/12/business/media-business-advertising-ads-are-weapon-battle-over-higher-auto-fuel-standards.html?pagewanted=all

12 Newell, *Climate for Change*.

13 Harriet Bulkeley, "No Regrets? Economy and Environment in Australia's Domestic Climate Change Policy Process," *Global Environmental Change: Human and Policy Dimensions* 11, no. 2 (2001): 155–169.

14 Article 4 of the UNFCCC (1992), http://unfccc.int/resource/docs/convkp/conveng.pdf

15 Robert Falkner, "Private Environmental Governance and International Relations: Exploring the Links," *Global Environmental Politics* 3, no. 2 (2003): 72–88; Philipp Pattberg, *Private Institutions and Global Governance: the New Politics of Environmental Sustainability* (Cheltenham, UK: Edward Elgar, 2007).

16 David Levy and Peter Newell, "Oceans Apart? Comparing Business Responses to the Environment in Europe and North America," *Environment* 42, no. 9 (2000): 8–20; Ian Rowlands, "Beauty and the Beast? BP's and Exxon's Positions on Global Climate Change," *Environment and Planning C: Government and Policy* 18, no. 3 (2000): 339–54; Ingvild Andreassen Sæverud and Jon Birger Skjærseth, "Oil Companies and Climate Change: Inconsistencies Between Strategy Formulation and Implementation?" *Global Environmental Politics* 7, no. 3 (2007): 42–63.

17 GE Eco-Imagination Report, *Investing and Delivering on Eco-imagination*, http://ge.ecomagination.com/site/downloads/news/2007ecoreport.pdf

18 The Bali communiqué on climate change, www.princeofwales.gov.uk/content/documents/Bali%20Communique.pdf

19 Eco-securities: company history, www.ecosecurities.com/Home/EcoSecurities-the_carbon_market/Company_history/default.aspx

20 Katherine Hamilton, Milo Sjardin, Thomas Marcello, and Gordon Xu, "Forging a Frontier: State of the Voluntary Carbon Markets," in *Ecosystem Marketplace and New Carbon Finance* (2008), www.ecosystemmarketplace.com/documents/cms_documents/2008_StateofVoluntaryCarbonMarket2.pdf

21 This paragraph draws from Peter Newell and Matthew Paterson, *Climate Capitalism* (Cambridge: Cambridge University Press, 2010).

22 CCX: Chicago Climate Exchange, *Overview*, www.chicagoclimatex.com/content.jsf?id=821

23 Beyond this, there are five other categories of participation that imply different levels of involvement: associate members, offset providers and aggregators, liquidity providers, and exchange participants.

24 Larry Lohmann, "Making and Marketing Carbon Dumps: Commodification, Calculation and Counterfactuals in Climate Change Mitigation," *Science as Culture* 14, no. 3 (2005): 203–235; Larry Lohmann, "Carbon Trading: A Critical Conversation," 2007, www.thecornerhouse.org.uk/pdf/document/carbonDDlow.pdf

25 Hamilton et al., "Forging a Frontier: State of the Voluntary Carbon Markets."

26 Heather Lovell, "Conceptualizing Climate Governance Beyond the International Regime: The Case of Carbon Offset Organizations," draft Tyndall Paper, May 2008.

27 Conversation with Johannes Ehberling of Eco-Securities, Oxford, 28 February 2008.

28 CDM Gold Standard (2008), www.cdmgoldstandard.org/objectives.php

29 Adam Bumpus and Diana Liverman, "Accumulation by Decarbonization and the Governance of Carbon Offsets," *Economic Geography* 84, no. 2 (2008): 127–55.

30 Voluntary Carbon Standard, "Voluntary Carbon Standard Program Guidelines," www.v-c-s.org/docs/Voluntary%20Carbon%20Standard%20Program%20Guidelines%202007_1.pdf

31 Voluntary Carbon Standard, "Voluntary Carbon Standard Program Guidelines."

32 The Climate Group, 2006.

33 Ecosystem Marketplace and New Carbon Finance, *State of the Voluntary Carbon Market* (2007).

34 Anja Kollmuss, Helge Zink, and Clifford Polycarp, *Making Sense of the Voluntary Carbon Market: A Comparison of Carbon Offset Standards* (Germany: WWF, 2008).

35 The Climate, Community, and Biodiversity Association, *Climate, Community and Biodiversity Project Design Standards (first edition)* (Washington, DC: The Climate, Community, and Biodiversity Association, 2005), www.climate-standards.org

36 E-mail received from Joanna Durbin, director, Climate, Community and Biodiversity Alliance, 30 July 2008.

37 Carbon Disclosure Project (2007), www.cdproject.net

38 The Greenhouse Gas Protocal Initiative homepage, www.ghgprotocol.org/

39 World Resources Institute (2007) www.wri.org/press/2007/12/iso-wri-and-wbcsd-announce-cooperation-greenhouse-gas-accounting-and-verification

40 Carbon Fund, "Carbon Footprint Protocol," 2008, 3, www.carbonfund.org

41 Simon Zadek, *The Civil Corporation: The New Economy of Corporate Citizenship* (London: Earthscan, 2001).

42 Peter Newell, "Managing Multinationals: The Governance of Investment for the Environment," *Journal of International Development* 13, no. 7 (2001): 907–19.

43 Michael Mason, *The New Accountability: Environmental Responsibility Across Borders* (London: Earthscan, 2005), 151.

44 Mason, *The New Accountability*, 150.

45 Peter Newell, "Civil Society, Corporate Accountability and the Politics of Climate Change," *Global Environmental Politics* 8, no. 3 (2008): 124–55.

46 ClimateChangeCorp, "Is Carbon Trading the Most Cost-Effective Way to Reduce Emissions?" *Climate News for Business*, www.climatechangecorp. com/content.asp?ContentID=6064

47 Newell, "Managing Multinationals," 907–19.

48 Robert Falkner, "Private Environmental Governance and International Relations: Exploring the Links," *Global Environmental Politics* 3, no. 2 (2003): 72–88; Lars Gulbrandsen, "Accountability Arrangements in Non-State Standards Organizations: Instrumental Design and Imitation," *Organization* 15, no. 4 (2008): 563–83.

49 Peter Newell and Matthew Paterson, "The Politics of the Carbon Economy," in *The Politics of Climate Change: A Survey*, ed. Max Boykoff (London: Routledge, 2009), 80–99.

6 Conclusions

1 Back issues of *Earth Negotiations Bulletin* can be found at: www.iisd.ca/ voltoc.html

2 Aaron Cosbey, *Border Carbon Adjustment* (Winnipeg, Canada: International Institute for Sustainable Development, 2008).

3 David Levy, "Business and the Evolution of the Climate Regime: The Dynamics of Corporate Strategies," in *The Business of Global Environmental Governance*, eds David Levy and Peter Newell (Cambridge, Mass.: MIT Press, 2005), 73–105.

4 Michael Lipskey, *Street-level Bureaucracy* (New York: Russell Sage Foundation, 1983).

5 The World Mayors and Local Governments Climate Protection Agreement, www.globalclimateagreement.org/

6 Heidi Bachram, "Climate Fraud and Carbon Colonialism: The New Trade in Greenhouse Gases," *Capitalism, Nature, Socialism* 15, no. 4 (2004): 10–12.

7 Bas Arts, "Non-State Actors in Global Environmental Governance: New Arrangements Beyond the State," in *New Modes of Governance in the Global System: Exploring Publicness, Delegation and Inclusiveness*, eds Mathias Koenig-Archibugi and Michael Zürn (Basingstoke, UK: Palgrave Macmillan, 2006): 177–201; Peter Newell, *Climate for Change: Non-State Actors and the Global Politics of the Greenhouse* (Cambridge: Cambridge University Press, 2000); Michele Betsill and Elizabeth Correll, "NGO Influence in International Environmental Negotiations: A Framework for Analysis," *Global Environmental Politics* 1, no. 4 (2001): 65–85.

8 Harriet Bulkeley, "Reconfiguring Environmental Governance: Towards a Politics of Scales and Networks," *Political Geography* 24, no. 8 (2005): 875–902.

9 Karin Backstränd and Charlotte Streck, "New Partnerships in Global Environmental Policy: The Clean Development Mechanism," *Journal of Environment and Development* 13, no. 3 (2004): 295–322.

10 Kevin Gallagher, ed., *Putting Development First* (London: Zed Books, 2005).

11 Peter Newell, "Civil Society, Corporate Accountability and the Politics of Climate Change," *Global Environmental Politics* 8, no. 3 (2008): 124–55.

12 Susan Strange, "Cave! Hic Dragones: A Critique of Regime Analysis," *International Organization* 36, no. 2 (1983): 488–90.

13 James Rosenau and Ernst Otto Czempiel, eds, *Governance Without Government: Order and Change in World Politics* (Cambridge: Cambridge University Press, 1992); Vasudha Chhotray and Gerry Stoker, *Governance Theory: A Cross-Disciplinary Approach* (Basingstoke, UK: Palgrave Macmillan, 2008); Rod A. W. Rhodes, "The New Governance: Governing Without Government," *Political Studies* 44, no. 4 (1996): 652–67.

Select bibliography

For more on theoretical approaches to climate governance and environmental governance in general (Chapter 1), see:

Harriet Bulkeley, "Reconfiguring Environmental Governance: Towards a Politics of Scales and Networks," *Political Geography* 24, no. 8 (2005): 875–902. Discusses the ways in which assumptions about scale and space in political analyses affect our understanding of the processes of governance.

Peter Newell, *Climate for Change: Non-state Actors and the Global Politics of the Greenhouse* (Cambridge: Cambridge University Press, 2000). Provides an assessment and explanation of the political influence of a range of non-state actors (the scientific community, mass media, business groups and environmental NGOs) upon the climate regime up to Kyoto.

———"The Political Economy of Global Environmental Governance," *Review of International Studies* 34 (July 2008): 507–29. Constructs a political economy framework for addressing key questions in environmental governance about who governs and how, what is to be governed (and what is not) and on whose behalf.

Chukwumerije Okereke, Harriet Bulkeley, and Heike Schroeder, "Conceptualizing Climate Governance Beyond the International Regime," *Global Environmental Politics* 9, no. 1 (2009): 58–78. Examines the strengths and weaknesses of regime approaches and global governance perspectives in understanding the governance of climate change, and assesses the potential contribution of neo-Gramscian and Foucauldian theory for such an analysis.

Matthew Paterson, *Global Warming, Global Politics* (London: Routledge, 1996). Assesses different strands of International Relations theory in terms of their ability to explain processes of international cooperation around the UNFCCC.

For more on the history of the climate negotiations and the two agreements which underpin the regime so far (Chapter 2), see:

Michael Grubb with Christian Vrolijk and Duncan Brack, *The Kyoto Protocol: A Guide and Assessment* (London: RIIA, 1999). Provides a useful assessment of the key features of the Kyoto Protocol and the negotiations that preceded it.

Irving Mintzer and J. Amber Leonard, eds, *Negotiating Climate Change: The Inside Story of the Rio Convention* (Cambridge: Cambridge University Press, 1994). Highlights the perspectives of leading state and non-state actors on the negotiations that led to the UNFCCC.

For more on the North–South politics of climate change and issues of equity (Chapter 3), see:

Neil Adger, Jouni Paavola, Saleem Huq, and M. J. Mace, eds, *Fairness in Adaptation to Climate Change* (Cambridge, Mass.: MIT Press, 2006). Explores the importance of issues of equity, justice and fairness in adapting to the effects of climate change.

Timmons Roberts and Bradley Parks, *Climate of Injustice: Global Inequality, North–South Politics and Climate Policy* (Cambridge, Mass.: MIT Press, 2007). Shows how North–South conflicts in the climate negotiations are a feature of broader historical inequities between developed and less-developed countries.

UNDP, *Human Development Report: Fighting Climate Change—Human Solidarity in a Divided World* (New York: Oxford University Press, 2007/8). Takes a systematic look at the developmental aspects of efforts to tackle climate change, with a strong emphasis on issues of rights, justice and historical responsibility.

For more on transnational governance networks and the role of community in climate governance (Chapters 4 and 5), see:

Liliana Andonova, Michele Betsill, and Harriet Bulkeley, "Transnational Climate Governance," *Global Environmental Politics* 9, no. 2 (May 2009): 52–73. Examines the origins and nature of transnational governance in relation to climate change and develops a typology for analyzing transnational governance initiatives.

Michele Betsill and Harriet Bulkeley, "Transnational Networks and Global Environmental Governance: The Cities for Climate Protection Program," *International Studies Quarterly* 48, no. 2 (2004): 471–93. Provides a detailed analysis of one example of a transnational governance network, Cities for Climate Protection, and the drivers and barriers to governing transnationally.

Gordon Walker and Patrick Devine-Wright, "Community Renewable Energy: What Should It Mean?" *Energy Policy* 36, no. 2 (2008): 498–99. Examines the ways in which communities have been mobilized to participate in climate governance through renewable energy projects.

For more on business responses to climate change and ways of understanding the role of business as an actor in climate governance (Chapter 6) see:

Kathryn Begg, Frans van der Woerd, and David Levy, eds, *The Business of Climate Change* (Sheffield, UK: Greenleaf Publishing, 2005). Provides a range of sectoral and country-based case studies of business responses to climate change.

Peter Newell, "Civil Society, Corporate Accountability and the Politics of Climate Change," *Global Environmental Politics* 8, no. 3 (2008): 124–55. Assesses the nature and effectiveness of a range of strategies adopted by civil society actors aimed at holding leading corporations to account for their responsibilities to act on climate change.

Simone Pulver, "Making Sense of Corporate Environmentalism: An Environmental Contestation Approach to Analyzing the Causes and Consequences of the Climate Change Policy Split in the Oil Industry," *Organization and Environment* 20, no. 1 (2007): 44–83. Looks at the origins of divergent corporate strategies in the oil industry with respect to climate change, and the implications of this for business responses to the issue.

Index

GLOBAL INSTITUTIONS SERIES

NEW TITLE
Preventive Human Rights Strategies

Bertrand G. Ramcharan, Geneva Graduate Institute of International and Development Studies

This book identifies the need for preventive human rights strategies, maps what exists by way of such strategies at the present time, and offers policy options to deal with the world of the future.

Selected contents: 1 Threats, challenges, and the responsibility to prevent 2 Obligations to prevent under international human rights treaties 3 The preventive role of national human rights institutions 4 Regional preventive strategies 5 Global preventive strategies 6 Preventive human rights diplomacy 7 The preventive role of peacekeepers, observers, and human rights monitors 8 Preventive strategies of NGOs 9 The preventive role of the international criminal tribunals and the International Criminal Court 10 Conclusion

February 2010: 216x138: 176pp
Hb: 978-0-415-54855-7
Pb: 978-0-415-54856-4
Eb: 978-0-203-85650-5

NEW TITLE
The UN Secretary-General and Secretariat
Second edition

Leon Gordenker, Princeton University

The new edition of this accessible introduction to the important role of the United Nations Secretary-General continues to offer a keen insight into the United Nations—the Secretariat and its head, the Secretary-General, summing up the history, structure, strengths and weaknesses, and continuing operations of an ever-present global institution.

Written by a recognized authority on the subject, this book continues to be the ideal interpretative introduction for students of the UN, international organizations, and global governance.

Selected contents: 1 Introduction 2 Blueprint and evolution of an international office 3 The UN Secretariat and its responsible chief 4 The Secretary-General as world constable 5 Promoting global general welfare 6 Reaching out to broader publics 7 Conclusion

February 2010: 216x138: 152pp
Hb: 978-0-415-77840-4
Pb: 978-0-415-77841-1
Eb: 978-0-203-85748-9

Routledge
Taylor & Francis Group

To order any of these titles
Call: +44 (0) 1235 400400
Email: book.orders@routledge.co.uk

For further information visit:
www.routledge.com/politics

GLOBAL INSTITUTIONS SERIES

NEW TITLE
Multilateral Counter-Terrorism

Peter Romaniuk, John Jay College of Criminal Justice, CUNY

Contemporary terrorism is a global phenomenon requiring a globalized response. In this book Peter Romaniuk aims to assess to what extent states seek multilateral responses to the threats they face from terrorists. *Multilateral Counter-Terrorism* is an essential resource for all students and scholars of international politics, criminology, and terrorism studies.

Selected contents: Introduction 1 Historical precedents for multilateral counter-terrorism: anti-anarchist cooperation and the League of Nations 2 Multilateral counter-terrorism and the United Nations, 1945–2001 3 Multilateral counter-terrorism and the United Nations after 9/11 4 Multilateral counter-terrorism beyond the UN 5 Multilateral counter-terrorism: today and tomorrow

March 2010: 216x138: 240pp
Hb: 978-0-415-77647-9
Pb: 978-0-415-77648-6
Eb: 978-0-203-85741-0

NEW TITLE
Global Governance, Poverty and Inequality

Edited by
Jennifer Clapp, University of Waterloo and
Rorden Wilkinson, University of Manchester

This book offers answers to questions raised about the role of global governance in the attenuation and amelioration of world poverty and inequality. The contributors interrogate the role of systems of governance at a time of global economic crisis and continuing environmental degradation against a backdrop of acceleration in inequalities within and between communities and across the globe.

Selected contents: Part 1 Development and the governance of poverty and inequality Part 2 Bretton Woods and the amelioration of poverty and inequality Part 3 Promising poverty reduction, governing indebtedness Part 4 Complex multilateralism, public-private partnerships and global business Part 5 Horizontal inequalities and faith institutions

May 2010: 216x138: 320pp
Hb: 978-0-415-78048-3
Pb: 978-0-415-78049-0
Eb: 978-0-203-85213-2

Routledge
Taylor & Francis Group

To order any of these titles
Call: +44 (0) 1235 400400
Email: book.orders@routledge.co.uk

For further information visit:
www.routledge.com/politics